LOST
SCIENCE

Astonishing Tales of Forgotten Genius

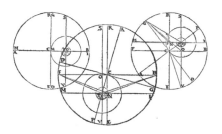

KITTY
FERGUSON

STERLING
New York

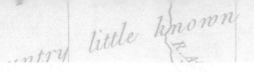

STERLING
New York

An Imprint of Sterling Publishing Co., Inc.
1166 Avenue of the Americas
New York, NY 10036

Text © 2017 by Kitty Ferguson

ISBN 978-1-4549-1807-3

Distributed in Canada by Sterling Publishing Co., Inc.
c/o Canadian Manda Group, 664 Annette Street
Toronto, Ontario, Canada M6S 2C8
Distributed in the United Kingdom by GMC Distribution Services
Castle Place, 166 High Street, Lewes, East Sussex, England BN7 1XU
Distributed in Australia by NewSouth Books
45 Beach Street, Coogee, NSW 2034, Australia

For information about custom editions, special sales, and premium and
corporate purchases, please contact Sterling Special Sales at 800-805-5489
or specialsales@sterlingpublishing.com.

Manufactured in Canada

www.sterlingpublishing.com

2 4 6 8 10 9 7 5 3 1

Design by Gavin Motnyk

For image credits, see page 326.

CONTENTS

———— ♦ ————

P. Adam Schall Germanus I. Ordinis Mandarinus

PREFACE

"Lost stories from the history of science. . . ." A brilliant idea, but if the stories are lost, what makes anyone think I might know them? That was my first reaction when the invitation came to write this book. Second thoughts prevailed: What ingenious sleuthing I could employ hunting for clues in obscure byways of science history. What long hours I could spend in library, university, and museum archives examining books and manuscripts no one had opened for years. What interesting, eccentric people I might encounter in abstruse areas of science. What spellbinding dinner conversations could be enjoyed with scientist and historian friends!

Fortunately, those friends *were* able to suggest a number of half-remembered stories they had stumbled across in their years as students and academics. And there the trails began. The same dinner companions raised a question: If *we* know about it, does it really qualify as "unknown"? How lost is "lost"? In order to be recoverable, a story surely has to be recorded in a book, or a letter, or the memory of at least one living person. Perhaps only a few people in the world know this tale, or perhaps only experts in one field of science vaguely remember it. That might legitimately qualify it as "unknown." Other possibilities required more of a judgment call: stories familiar to those in one branch of science, not at all to those in another, and even less to the general reading public; stories of out-of-the-mainstream work of people whose names are household words for other celebrated discoveries.

It was frustrating to find that some stories must remain lost—at least until a more skilled sleuth comes along or sheer luck turns up more information about them. For example, take Francesco Giuntini, a Florentine astronomer and astrologer who wrote a vast tome titled *Speculum*

A portrait of Johann Adam Schall von Bell (1592-1666), a German Jesuit missionary and astronomer, in China (see page 9). From La Chine d'Athanase Kirchere de la Compagnie de Jesus, 1649, by German Jesuit scholar Athanasius Kircher.

Astrologiae (which still resides in the Bodleian Library at Oxford). He erroneously predicted horrific consequences to follow the comet of 1577, and he died at Lyons in 1590, "killed by his own books." How? The explanation is probably prosaic: perhaps a copy of *Speculum Astrologiae* fell on his head. But this cryptic quotation from an unknown biographer suggests a range of intriguing possibilities and conjures up visions of that book and others suddenly turning against him with murderous intent. A fiction writer will have to take it from there.

The first chapters in this book are personal favorites: outrageously adventurous tales, riveting because the colorful people in them are irrepressible risk-takers, spies, scoundrels, and heroes: French astronomer Jean Chappe d'Auteroche, whose scientific curiosity took him to his death; American-born Benjamin Thompson (aka Count Rumford), an exceptionally skilled inventor and scientist but ruthless about using others to his own advantage; Flemish Jesuit astronomer Ferdinand Verbiest, who so cleverly used his science to save his life. Did Mark Twain know about him when he wrote *A Connecticut Yankee in King Arthur's Court*? Verbiest's adventure took place in an even more exotic setting: the seventeenth-century imperial court in Beijing.

My second category features people who were essential to major advances but failed to receive the credit they deserved, who are lost in the shadows of celebrated colleagues—or, more curiously, lost in their own shadows. Some of the best examples are too well known to qualify as "lost." I decided not to include Rosalind Franklin, who most agree should have won the 1962 Nobel Prize in Physiology or Medicine with Watson, Crick, and Wilkins for the double helix of DNA; Jocelyn Bell Burnell, who watched her supervisor Anthony Hewish receive the Prize in Physics despite being the first to observe pulsars; or Milton Humason, who may have been more responsible than he ever claimed for Edwin Hubble's groundbreaking discoveries. I chose instead to write about the Austrian physicist Lise Meitner, who is sadly unfamiliar to most aside from those interested in the history of their own field of atomic research and a few among my friends in Cambridge, England, who are old enough to have known her personally. She certainly should have shared the 1944 Nobel Prize in Chemistry, which was awarded to her colleague Otto Hahn, for splitting the atom. Probably the least unknown of my subjects is British naturalist and explorer Alfred Russel Wallace. Should he be feted along with, or even ahead of, Charles Darwin for the discovery of natural selection? Wallace himself certainly thought *not*! Mary the Jewess (the most ancient of my stories) and Maria Sibylla Merian are both remembered in ways that neglect what they really achieved. They are lost, as it were, in their own legends.

My third group of stories is about achievements, discoveries, and lines of discovery that many of us might recognize while having no idea who actually got there *first*. Who wrote the first science fiction book (and risked being executed because of it)? Who worked out mathematically how Earth's orbital movements affect the coming and going of ice ages? Who discovered neurofeedback and deserves to be called the "Pavlov of Cats"?

Science continually builds on itself, as political and social history do not always do. Significant discoveries generally lead to other significant discoveries; they are not dead ends or isolated. Some "lost" stories of science turn out to be far from this mainstream and would arguably deserve oblivion were it were not for a side story, adventure, or personality that makes them especially interesting. They are *good stories*: The canny way Ferdinand Verbiest used his scientific skills to save himself and his colleagues in Beijing and the little automobile he invented to amuse the emperor of China make it possible to forgive him his regrettable adherence to Tychonic astronomy. Count Rumford's invisible ink, spying for the Redcoats, and innovative kitchen and fireplace designs upstage his barely remembered threat to the caloric theory of heat. Some of the stories are wonderful travelogues, taking their protagonists to unexplored, dangerous parts of the world. I have found interesting material in diverse branches of science and pre-science and from many periods of history—and the subjects of these stories include both men and women.

Some readers of this book will exclaim, "Why didn't you include [insert name of overlooked scientist here]?" You have my apologies. If you send me your suggestions, perhaps I can compile a second volume of obscure tales. I know some of those other stories, too, and wish there had been room to include them.

Researching this book has involved some extraordinarily helpful, imaginative people with good memories. My greatest debt of thanks is to Melanie Madden, former editor at Sterling Publishing, who came up with the idea for the book and invited me to write it. Also at Sterling, I am thankful to Kayla Overbey, production editor; Gavin Motnyk, senior designer; Barbara Berger, executive editor; and David Ter-Avanesyan, cover designer. The scientists and other academic friends who led me to stories are Nick Shackleton for Milutin Milanković, P. Susie Maloney for Benjamin Thompson/Count Rumford, Owen Gingerich for Jean-Baptiste Chappe d'Auteroche, Joseph Sandford for Barry Sterman, and Robert G. W. Anderson for a host of excellent leads. There were those who provided help when the trail seemed to reach an insurmountable impasse: Michael Loewe at Clare Hall in Cambridge advised me on the proper ways to identify Chinese emperors and told me that I needn't go to Belgium or Beijing to uncover information about Ferdinand Verbiest; the archives of the East Asian History of Science Library at the Needham Research Institute were just around the corner. Royal Society archivist Katherine Harrington extricated, photocopied, and e-mailed me a missing page from the 1703 *Philosophical Transactions of the Royal Society* that had somehow been skipped when the records were digitized. André Berger at the Université Catholique de Louvain made it possible for me to use an otherwise unavailable autobiography of Milutin Milanković from his personal collection and brought me up to date on his own work on the astronomical theory of the ice ages. Tom Graedel at Yale University cleared up a definition glitch and disagreed with the Internet's insistence that "snow budget" meant money allocated for snow removal. The staff at the Archives Centre at Churchill College in Cambridge were endlessly patient, bringing out file after file of Lise Meitner's correspondence. A librarian in the Aoi Pavilion of the Cambridge University Library located a truly unfindable book that had been a little too creatively cata-logued in that enormous library. Kate Heard, senior curator of prints and drawings for the Royal Collection Trust at Windsor Castle, pointed me to the one good translation of Maria Sibylla Merian's book about her sojourn in Surinam. Zbigniew Wesolowski, SVD, at the Institute Monumenta Serica in Sankt Augustin, Germany, gave me permission to use Ditlev Scheel's drawing of Verbiest's "automotive machine." Barry Sterman brought me up to date on his most recent findings about the brain mechanisms involved in neurofeedback. Nick Barrand gave me invalu-able help with last-minute corrections to captions in my chapter about Milutin Milanković. Librarians and staff also helped me at the Cambridge University Library Rare Books Reading Room, the New York Public Library Dorot Jewish Division, the NYPL Science, Industry and Business Library, the Needham Research Institute in Cambridge, the J. Paul Getty Museum in Los Angeles, and the Bluffton Public Library in South Carolina.

I must not end this preface without expressing my heartfelt thanks also to my two splendid literary agents: Meg Davis in London and Rita Rosenkranz in New York.

I am grateful to you all.

Kitty Ferguson

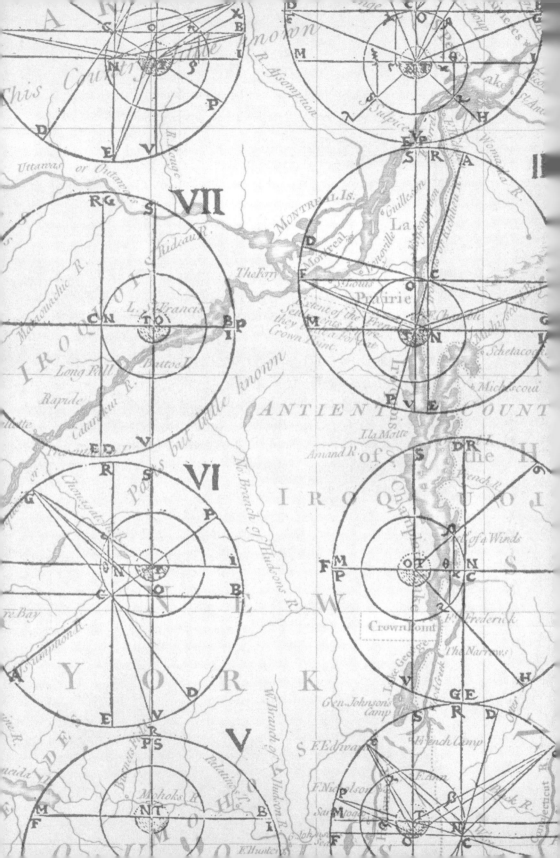

PART I

RIPPING YARNS

ONE

Ferdinand Verbiest

THE EMPEROR'S NEW ASTRONOMY

(1601–1688)

N 1668, THE "SEVENTH YEAR OF KANGXI," the emperor of China was a powerless teenager, fourteen years old. His father, on his deathbed, had appointed four men to rule as regents until his son came of age. He had made unfortunate choices. Deadly clashes among the four regents had left two survivors—dangerous, ruthless men determined to keep the young heir to the throne under their control as long as possible. The boy was well educated and extremely intelligent. At age thirteen he had been allowed to attend to affairs of state, though ruling only in name. He had nevertheless contrived to acquire a fearfully good grasp of the politics of the Chinese court and the workings of the Empire. The time was approaching when, according to law and tradition, he could dispense with the regency, and he was chomping at the bit. With the advice

possibly of his tutor and his mother, the dowager empress, he looked for a way to rid himself of the two surviving regents. His choice of weapon sounds curiously innocuous: the calendar. It was, in fact, an astute choice, and this teenager was about to prove that he, too, could be a

OPPOSITE: *Jesuit astronomers meet with the Kangxi Emperor in an imaginative tapestry from* The Story of the Emperor of China Series, *woven in France, c. 1700.* LEFT: *A Chinese scroll portrait of the Kangxi Emperor at a calligraphy table, c. 1662–1722.*

dangerous man. He was the "Kangxi Emperor," second ruler in the Qing Dynasty and destined to be one of the ablest monarchs in China's history.

While historians and scientists as well as non-experts are familiar with the story of Galileo's trial in Rome in 1633, hardly anyone knows about a trial that took place in the young Kangxi Emperor's court in Beijing thirty-five years later—one that pitted religion against religion, science against science, culture against culture. The outcome depended on clear-cut scientific results that all present could agree upon, though few could understand. Lives and the control of an empire hung in the balance, but it was the calendar that lay at the root of the contest. The earliest first-person description of the event was lost for three hundred years until 1991, when historian of science Efthymios Nicolaidis discovered a manuscript in a collection in the Church of the Holy Sepulchre in Istanbul, referenced sketchily in an old catalogue. He suspected he had found a previously unknown work from the seventeenth-century Jesuit astronomer Ferdinand Verbiest. Noël Golvers, a specialist in the early Christian missions in China at the Ferdinand Verbiest Foundation in Leuven, Belgium, was able to certify that this manuscript was, indeed, unknown. Dating from 1676, it was the earliest Latin text from Verbiest. It described in vivid detail Verbiest's unusual experiences in the imperial court in Beijing.

The story had begun in 1601, a half century before the Kangxi Emperor's birth, when Matteo Ricci, Jesuit missionary and astronomer, settled in the Chinese capital and learned that the reigning Wanli Emperor had a penchant for European mechanical instruments such as clocks and harpsichords. Here was an opportunity to gain access to the highest strata of Chinese mandarins and others who frequented the imperial court, even to the emperor himself. Ricci had a particularly valuable card up his sleeve: Western mathematics and astronomy, and their potential for helping Chinese astronomers produce an accurate calendar. The European *Prutenic Tables* of 1551—based on Copernican sun-centered astronomy and the more advanced computational methods of Europe—were unmatched by anything available to Chinese scholars at the time. At best, these scholars could come no closer than about a quarter of an hour in predicting an eclipse—an essential step in producing the calendar. Ricci and his associates could do so dependably with a discrepancy of no more than one minute.

The Jesuits were devout missionaries with a commission to preach the Christian gospel to all nations, and they strategized that if they were to Christianize China, it was at the pinnacle of Chinese society, Beijing's imperial court, that they must sow the seeds of their faith. Wisely, Ricci insisted that candidates for the Chinese mission field possess not only

A portrait engraving of Verbiest from Description de la Chine *(The* General History of China*), 1736, by French Jesuit and historian Jean-Baptiste Du Halde.*

high religious and moral standards and the psychological strength to cope with life in an alien culture, but also a strong mathematical background. Such men were not easy to find in European Jesuit universities, where the mathematical sciences had fallen into somewhat low esteem. Nevertheless, fine scholar-missionaries did follow Ricci to China in the first half of the seventeenth century. Among their early converts to Christianity was an impressively able Chinese astronomer, Xu Guangqi. Together with him, the Jesuits established an Academy for Western Mathematical Sciences in Beijing in 1629, built astronomical instruments, and published an immense "astronomical encyclopedia" consisting of around 120 chapters.

China's rich tradition of astronomy had begun before the dawn of written history. Ancient observations there surpassed anything in Europe in their time. Even in the early seventeenth century when Ricci arrived, Chinese astronomy was in some respects more enlightened than astronomy in Europe, where most scholars still clung to Aristotle's and Ptolemy's Earth-centered model in which the stars are all equidistant from Earth, and the planets and sun orbit Earth in fixed, concentric, "crystalline" spheres. Chinese scholars, by contrast, had decided two millennia earlier that the heavenly bodies float in infinite empty space. They'd also known of sun spots for many centuries—long before their discovery in the West.

Much Chinese astronomical and mathematical expertise over the millennia had been devoted to producing yearly calendars. According to tradition, sometime before 2000 BCE the emperor Yao had ordered that the calendar be regulated so that celestial events could be predicted. An accurate calendar was essential to the empire's activities and the functioning of its mammoth bureaucracy, but it also had a ceremonial, religious significance that is inconceivable to those of us in the modern West. The emperors of China were called by the title "Son of Heaven," signifying a special relationship with the skies above. If a man with such a title didn't know when the next eclipse would occur, then there was reason to doubt his right to the throne. Considerable understanding of astronomy and computational skill were focused on the correct prediction of celestial events, particularly

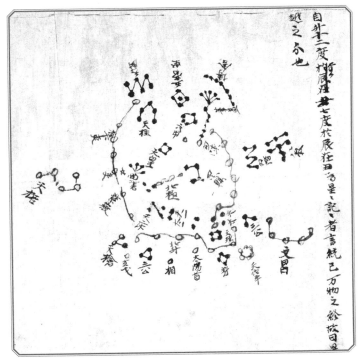

The Dunhuang Star Atlas, a remarkably accurate Chinese astrological chart dating to c. 700 CE, is the oldest known completely preserved star atlas ever discovered.

eclipses of the sun and moon, and this precious expertise passed from generation to generation. To further ensure the calendar's accuracy, probably as early as 25 to 220 CE, Chinese emperors set up an institution that was unique in the world even as late as the seventeenth century. Its name translates into English as the Chinese Astronomical Bureau.

Unfortunately, the late Ming dynasty—in power until 1644—had allowed the Astronomical Bureau to languish in the hands of scholars less skilled than their forebears. With little fresh input, they were still largely dependent on observations made in the thirteenth century by Islamic and Chinese scholars connected with or strongly influenced by the great Persian observatory at Marāgha, or Maragheh, in what is now northwestern Iran. The Marāgha observations had been some of the finest of their era, but they were not completely free of small inaccuracies. At first these had been of no consequence in predicting celestial events. Even over time it had been possible to ignore or compensate for them, but over a span of years tiny discrepancies had grown to noticeable and eventually intolerable proportions. The heavens had, of course, continued to move in their own way, not necessarily in agreement with the best thinking and predictions of earthly experts. By the end of the fifteenth century, errors in early Marāgha astronomy were producing serious failures in prediction, particularly of eclipses.

The public manifestations were embarrassing. Festivals connected with astronomical events and seasons were falling at the wrong time. In 1592, the Bureau's prediction of a lunar eclipse was off by one whole day, and the "relief" festival, traditionally held following an eclipse to prevent the calamities it portended, had to be postponed. Discrepancies in the lunar calendar also meant discrepancies in predicting total eclipses of the sun, for only on the first day of the lunar cycle, with a new moon, is it possible for a total eclipse of the sun to occur. When a solar eclipse fell on the wrong day or failed to happen, it was a disaster, considered by many to be an extraordinarily dire portent, and certainly an embarrassment to the emperor. The Bureau soon realized that failing to predict an eclipse correctly and mis-scheduling the "relief" festival was a far more serious offense than predicting an eclipse that failed to occur, so they began over-predicting eclipses. When a predicted eclipse didn't happen, it could be taken as mercy linked to heavenly approval for the emperor. But such subterfuge could not work indefinitely, and an ancient tradition of scholarship was losing its credibility. The prestige of the "Son of Heaven" and the social and dynastic stability of the empire were in peril. The situation at the Astronomical Bureau was becoming desperate as, one after another, attempts at reform produced no improvement.

In 1644, the Manchus, led by the young Kangxi Emperor's father (the Shunzhi Emperor), toppled the Ming dynasty, which had ruled for seventy-six years. Upon establishing the Qing dynasty (which would rule until 1911/12), one of the Shunzhi Emperor's first acts was to declare the calendar produced by the Jesuit missionary Johann Adam Schall von Bell the official calendar of China. There were rumors that Schall helped to cure the dowager empress of a mysterious illness, and that this was the reason he became not only a scientific adviser but also a close friend of the Shunzhi Emperor, who called him *mafa* ("grandfather") and made him director of the Astronomical Bureau. The Jesuits—though not without some opposition and grumbling from native Chinese within the court—had reached their goal at the highest strata of imperial society and held on to it through a mammoth shift in political power. As they had hoped, the prestige and credibility of their astronomy were lending prestige and credibility to their religion. In 1650, the emperor allowed Schall to build a church in Beijing and, on occasion, even came to services.

A page from an Imperial Chinese astronomical calendar, c. 1523, used by the Chinese Astronomical Bureau during the Ming dynasty.

Thus things stood in 1658, when thirty-four-year-old Ferdinand Verbiest sailed from Lisbon to Macao with a fleet that made the long, treacherous voyage once a year. His credentials as an astronomer were not impressive: one semester under the tutelage of the mathematician André Tacquet at the University of Leuven in Belgium; ten undocumented years in Rome, where he *might* have undertaken private study with the polymath Athanasius Kircher; and a year teaching mathematics at a Jesuit school in Coimbra, Portugal (where, by his own admission, he "learned mathematics more than I taught"), were scant preparation for the challenges lying ahead of him. Once the fleet had embarked, however, Verbiest found it possible to ignore the perils of this two-year voyage, which had a history of fatal disease and shipwrecks, and concentrate on his beloved astronomy. Fellow passenger Martino Martini, an Italian Jesuit astronomer, was returning to the mission field. Verbiest commented in a letter to Kircher, "Often, in the sublunary night, we learnt from Father M. Martini not superstitious astrology, but the rules of astronomy and the rise of the stars."

Before making too hasty an assumption that the astronomy of Ricci, Schall, Martini, and Verbiest was state-of-the-art Copernican sun-centered astronomy, it is best to examine more closely what "state-of-the-art" meant back in Europe at the time. "Physics" at the University of Leuven, where Verbiest spent the one semester studying with the mathematician Tacquet, meant the ancient physics of Aristotle. Astronomy was almost entirely medieval astronomy based on the first-millennium work of Arabic-Ptolemaic astronomers al-Farghānī and al-Battānī. A century had passed since Copernicus had written his book *De revolutionibus orbium coelestium* (*On the Revolutions of the Heavenly Spheres*), and Verbiest may have heard some professors at Leuven discussing the systems of Copernicus and Tycho, but only as interesting hypotheses. That was the only nod given to emerging ideas.

These were the decades during which Copernican astronomy, after three quarters of a century's peaceful coexistence with Catholic authority, increasingly ran into difficulty as Galileo insisted on a showdown. Jesuit scholars

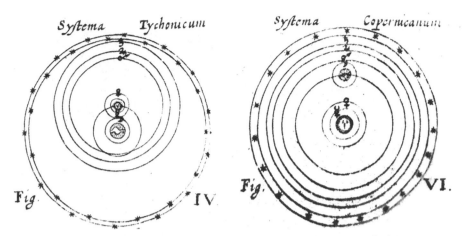

The Tychonic and Copernican cosmological systems from Athanasius Kircher's Iter Exstaticum (Ecstatic Journey) *(1655), which he wrote in the form of a celestial dream. In Tycho's model, Earth remains in the center of the system. The sun orbits Earth, and all the planets orbit the sun. In Copernicus's model, the sun is at the center of the system. All the planets, including Earth, orbit the sun. Although it is not clear in these early drawings, the Tychonic and Copernican systems are geometrically equivalent.*

in Europe found themselves between a rock and a hard place. Being fine mathematicians with a superb scholarly tradition and a passion for advancing technology, they could not deny the validity of the discoveries Galileo was making with a new instrument, the telescope—discoveries that supported Copernican sun-centered astronomy. On the other hand, they could not throw aside the theological traditions of their faith—extrapolated not directly from the Bible but from long-accepted interpretations of it and the authority of St. Augustine and other church fathers—which followed Aristotle in insisting that the cosmos was centered on an unmoving earth. Surprising to the many of us who are taught that, after a few glitches, Copernican astronomy quickly supplanted Ptolemaic and Tychonic astronomy, it in fact took many decades for the change to take place in Europe. In China, Copernican astronomy didn't gain a firm footing until Protestant missionaries arrived in the early 1800s.

Meanwhile, the great, eccentric Danish astronomer Tycho Brahe had thrown scholars like the Jesuits a lifeline by proposing an ingenious model that kept the earth as the unmoving center of everything, with the sun

An engraving from Tycho Brahe's book Astronomiae instauratae mechanica (Instruments for the Restoration of Astronomy), *1598, showing Brahe and his Great Mural Quadrant. The engraving is a reproduction of a wall painting—a scientific and artistic masterpiece— in his palace observatory in Uraniborg, Isle of Hven (now part of Sweden).*

orbiting the earth and all the other planets orbiting the sun. It was a brilliant compromise—the exact geometric equivalent of the Copernican system. Until Isaac Newton's discoveries about gravity in the late seventeenth century, there was little to argue the validity of one system over the other, except that one was decidedly simpler and more economical. Johannes Kepler had made adjustments when compiling his *Rudolphine Tables* in the light of his own reliance on Copernicus and his discovery that planetary orbits are elliptical, but those subtleties went largely undetected by those who used the tables and were not important for astronomers of the Chinese Astronomical Bureau in the day-to-day practice of computing and predicting eclipses. Thus, the *new* European astronomy that the Jesuits brought to China was not, in fact, sun-centered Copernican astronomy. It was the compromise astronomy of Tycho Brahe—a system more in accord with Jesuit teaching.

By the 1660s, Adam Schall von Bell was growing old in the happy knowledge that his time in China had been fruitful for him and his order. He began to consider who might succeed him in the high-ranking positions he had achieved. Ferdinand Verbiest was serving in a missionary post in a western province, but he already enjoyed high regard as a mathematician. Schall chose him as successor. In a letter supporting his choice, he wrote "according to his reputation, [Verbiest] is well versed in all kind of virtues and letters, but especially in mathematics." The Shunzhi Emperor summoned Verbiest to Beijing.

Verbiest's grooming by Schall was meticulous and anything but hasty. The elderly astronomer introduced him to the routine, practical duties he would have to perform, and Verbiest made an arduous effort to master the Chinese language. By 1663 he may have surpassed the expertise of his mentor. When a bell weighing 120,000 pounds was about to be hoisted into a tower, Verbiest calculated a way to increase the power and efficiency of the tackles. Schall heeded his advice and averted a disastrous failure. In the following year, Verbiest received the emperor's permission to construct six astronomical instruments, in the style of Tycho Brahe's magnificent pre-telescope instruments, for the Beijing Observatory. Everything seemed to be moving ahead toward a smooth transition.

An elaborate quadrant designed by Verbiest, from his book
Xinzhi yixiang tu, *published in Beijing in 1674.*

. A peaceful old age for Schall was not to be, however. It is from this point in the story that we have Verbiest's own account from the manuscript discovered in 1991.* The Shunzhi Emperor died, and the ill-chosen regents he had appointed to rule for his heir were deeply hostile to the Jesuit astronomers and to all European scholarship and religion. One virulent enemy at court—Yang Guangxian, who was something of an astronomer himself, though not very skilled at it—collaborated with Chinese astronomers who resented the success of Western astronomy and charged Schall with plots against the recently deceased emperor. Among the accusations leveled at

---

* It is difficult to imagine in our own day of rapid communication and travel how isolated Verbiest was in Beijing from the progress of science occurring in Europe. Long after the events he described had taken place, he decided to act on the conviction that closer contact would be advantageous not only for himself and his scientific and religious associates but also for the two cultures. He became involved in diplomatic efforts to open an overland route from Beijing to Europe by way of Russia and Poland, making overtures to Czar Alexey Mikhailovich of Russia and King John III Sobieski of Poland. The manuscript discovered in 1991 was part of Verbiest's appeal to the Czar, in the interest of awakening his enthusiasm for that route.

Schall were causing the Shunzhi Emperor's death by selecting an inauspicious date for his infant son's burial and insultingly describing the Chinese as unimportant descendants of the ancient Hebrews. The regents banned Catholicism, and the churches closed. The duty of compiling calendars was reassigned to native scholars who reverted to the older Chinese methods. Yang Guangxian became head of the Astronomical Bureau.

In the most extreme telling of the story—though not in Verbiest's more diplomatic account—Schall, Verbiest, and their associates were chained in a filthy prison cell, bound to wooden pegs in such a way that they could neither stand nor sit. Whatever the details of that imprisonment, the elderly Schall suffered a stroke and was left unable to speak. Verbiest, still not proficient in the Chinese language, made a pitiful and unsuccessful attempt to defend him at his trial. The sentence for Schall and his colleagues was death by strangulation, which a higher court changed to death by dismemberment. The prisoners were awaiting execution, about to become martyrs to their astronomy and their faith, when an earthquake destroyed the part of the palace where the dismemberment was to be carried out. Not surprisingly, the imperial court and even the regents could not completely dismiss the possibility that this was an unfavorable sign, and although five Chinese Christian astronomers were executed, the sentences for the European astronomers were changed to banishment to Canton. Only Verbiest, Schall, and two others—Lodovico Buglio from Sicily and Gabriel de Magalhães of Portugal—were retained in Beijing under house arrest. Buglio and de Magalhães were considerably older than Verbiest, both in their early sixties. Schall, elderly and extremely feeble, died during this lighter imprisonment. Verbiest not-too-unhappily devoted this enforced leisure time to his personal study of mathematics and astronomy.

Meanwhile, the Jesuits' accuser Yang Guangxian and his assistant Wu Mingxuan were becoming increasingly aware of the predicament their lack of expertise and inexperience were causing at the Astronomical Bureau. They petitioned their superiors for help in finding other experts, but either their plea was ignored, or none could be found. Yang and Wu went ahead and produced calendars, recognized that they contained discrepancies, and petitioned a second time for expert help. However, the season for the elaborate ceremonies for presentation of a new calendar was approaching too rapidly, and the Ministry of Rites decided that Wu's calendar, being closest to accurate, would be chosen.

It was four years into the Jesuits' imprisonment when the fourteen-year-old Kangxi Emperor, rightly suspicious about the accuracy of the current calendars and the failure to find any experts to aid the Bureau, began thinking he might use the calendar crisis to bring down the two regents. The most embarrassing problem that anyone could remember had arisen. The year was to have thirteen months rather than the normal twelve. It came to the young emperor's ears that there were, indeed, other experts available: European astronomers whom his father had respected, who had been in prison since shortly after his father's death. He sent emissaries to seek them out, if they were still alive, and ask their opinion of the calendars for the current year and the new year. Verbiest's reply was blunt: The calendars were "teeming with mistakes."

The Kangxi Emperor summoned Yang and Wu and two of their Chinese colleagues to the palace, along with Verbiest, Buglio, and de Magalhães, and ordered all of them to bury the hatchet and decide in a fair and impartial manner the best way to compute the calendar. The Jesuit astronomers were about to face a contest with Wu, whom Verbiest described as an "overbearing and unbridled man" and "that most criminal old man, Yang Guangzian [*sic*]," who was responsible for their trial, imprisonment, and death sentences. Verbiest knew that with a fair-minded emperor, this was a battle he was equipped to win. His description is full of the spectacle and suspense of the occasion, his own excitement, and his confidence that he could, with the help of God, save his own and his colleagues' lives.

The first royal summons was to a spacious hall within the palace walls, which must have been a disorienting, intimidating experience after four years in narrow confines. Yang and Wu were in the room, along with a crowd of distinguished mandarins who would serve as witnesses. A third powerful Chinese astronomer, Ko-Lao, remained in the background. The emperor chose not to make an appearance yet, and the two regents were not present. It was a shrewd guess on the part of the emperor that the regents would consider this matter peripheral to great affairs of state and beneath their notice.

Later that day, at a distance, the young Kangxi Emperor first met Verbiest, Buglio, and de Magalhães. The court was organized with mandarins of the Astronomical Bureau on the front row and Verbiest (the youngest in attendance, other than the emperor) in the last. In sequence, according to rank, the mandarins genuflected before the emperor, who gave orders that Verbiest move to the front where he could see him. He had never before met a European. Addressing Verbiest directly, he inquired whether there was any way to prove "before our eyes" whether the calculation of the calendar corresponded to the truth of what was happening in the heavens. Verbiest replied that there was: The sun being the "principal" and most easily seen of all the heavenly lights,

> I will immediately calculate, here and now, before Your Majesty's eyes, after whatever gnomon* has been erected or whatever table or seat has been placed in the middle of this courtyard, how long the shadow will be, which that gnomon will project at a determined hour. On the basis of the given length of this shadow, the sun's altitude at the same hour can manifestly be inferred, [and] from the height of the sun on a given day and hour, its place in the heaven; and so whether it corresponds with the celestial position found in the calculations of the calendar.

Verbiest recorded that this answer appeared to delight the emperor, while Verbiest's enemies looked as if the proceedings had turned "gloomy and fatal." Then the emperor turned to the Chinese astronomers and asked if they knew about this art of calculating shadows. All kept silent except Wu Mingxuan, who said he did and that it was a reliable test. The emperor went on to question Verbiest on various matters, as Verbiest recalled, "with an always-benevolent eye," and then declared that Wu, Verbiest, Buglio, and de Magalhães would conduct experiments at the Beijing Observatory

---

* The most familiar example of a gnomon is the projection on a sundial that casts the shadow.

to test their respective methods of computation. He dismissed the court and ordered Ko-Lao and the other mandarin astronomers to set up the gnomon. At this point, Wu got cold feet and confessed that, although he knew of shadow calculation, he knew nothing about how to do it. The emperor was indignant and sent him off to await the result of Verbiest's test and his own "severe punishment." At noon the next day, Verbiest alone would predict the length of a gnomon's shadow.

The gnomon Verbiest was assigned was an ancient bronze column—eight feet three inches (2.5 meters) tall—that over the centuries had tilted until it no longer stood at a right angle to the elaborate bronze table on which it was fixed. Overnight, Verbiest and his two associates hastened to compensate for this problem. On top of the column they added a wooden construction known as a horizontal cross lath, creating a surface that was perfectly horizontal and at the height assigned by the Chinese astronomers. From that, Verbiest dropped a plumb line to the table below. The point at which that line fell determined the point from which he counted out the fractions of shadow length to be measured, and he painted a line on the table to indicate where the edge of the shadow of the gnomon should be at noon the next day. The sun was close to the winter solstice, so the shadow would be long.

The next morning, the mandarins gathered at the observatory and watched as the sun climbed toward its highest point. "Behold," wrote Verbiest, "the shadow of the above mentioned gnomon reached very exactly the line which I had marked before on the table. This caused utter astonishment." Verbiest was a little astonished himself.

This result was reported to the emperor, who was very pleased and ordered another test the next day—this time nearer his own person, within a courtyard of the palace. The mandarins again were to set the height of the gnomon, and they chose the length of a brass ruler that Verbiest was carrying with him: 2.2 feet (sixty-seven cm). The Jesuits returned to their residence, and de Magalhães, who was adept at designing and constructing mechanical devices, again stayed up all night fabricating a perpendicular gnomon that

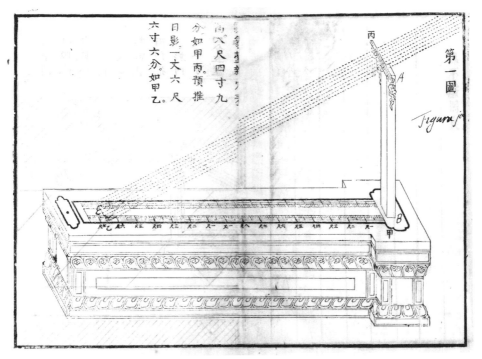

Verbiest's drawing of the First Gnomon Experiment, from his book Qin ding xin li ce yan ji lue (Astronomical Observations), *published in Beijing in 1669.*

stood 2.2 feet (sixty-seven cm) high, set on a horizontal table subdivided into feet and inches, with three set screws to aid in the calculation. Again, Verbiest drew a cross line where he calculated the noon shadow would end.

The next day, they brought the device to the palace and set it up in the courtyard. This time even more mandarins and some women of the imperial court sat in all their finery in a circle around the table and the gnomon. There were murmurs of admiration for the rapidity with which these had been constructed. Tension mounted as the sun approached its noon position. The gnomon's shadow had not yet fallen on the horizontal table. It appeared instead to waver on the flat floor of the courtyard and exceed the distance that Verbiest had marked. Verbiest's adversaries began to whisper and laugh and sneer among themselves.

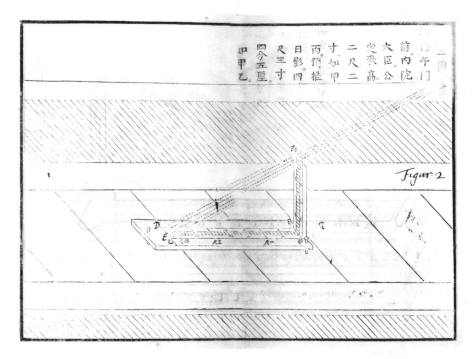

Verbiest's drawing of the Second Gnomon Experiment, from Qin ding xin li ce yan ji lue.

But behold . . . as the sun approached closer and closer its midday point, the shadow ascended my table, contracted suddenly its former length, and came very close to the transversal line I had drawn on the table. Finally when the sun reached its midday point, the shadow coincided perfectly with the line signed on the table. The Manchu mandarin, who was the most prominent amongst them, shouted perplexed: "We have here a great master, indeed!" My adversaries, on the contrary, glanced in turn at each other with pale faces.

It was a splendid victory, but one that would cast a long shadow that Verbiest compared to the inexorable shadow of a gnomon. He had made lifelong enemies.

For the moment, however, all was going extremely well for Verbiest and for the emperor, who asked that the table and gnomon be brought in so that he could see them. Again, delight and, again, orders for yet another demonstration—this time back at the observatory with the original bronze gnomon. The emperor explained that the settling of a matter of such immense importance should not be left open to any possible suspicion that it had been decided too casually or without due consideration. He was stacking his cards carefully. Verbiest and his two associates must have sighed as they returned with sinking hearts and renewed misgivings to their calculations and construction, for Verbiest admitted in his account that, in his private studies and experiments at home, his results had by no means always been so successful and unerring. He had almost always "observed some deficit or surplus between the shadow and my calculations."

Nevertheless, the next day, having adjusted the height of the new horizontal wooden cross lath to match the newly assigned height, and having again drawn the deciding transversal line on the bronze table, Verbiest stood back and watched with his fellow Jesuits and the same august company of mandarins as the tip of the shadow crept toward the noon point and touched exactly the line Verbiest had drawn. Verbiest went on in his account to say that in all fairness, the success of these tests could only be ascribed to the work of God, "surely something which entirely exceeded my diligence and assiduity!" His own calculations and the mounting of his instruments were, frankly, not that precise.

The emperor, now convinced that Verbiest had extraordinary skills, immediately sent him two calendar books to examine. Wu Mingxuan had been the most recent to have a shot at correcting the calendar and had produced these books by mixing Chinese and Arabic astronomy in what Verbiest called a "Chinese-Arabic" calendar. Verbiest listed the most serious problems in this work and presented them in a "petition" to the emperor, who summoned to the palace the imperial princes (his own kinsmen), the grandees of the first class, and all the most important mandarins of all orders and ministries for a ceremonial General Meeting of the Official Council of Deliberative Officials of the Empire. Clearly the calendar was of grave importance, but this gathering was also a key moment in the young emperor's power move.

The General Meeting read Verbiest's petition and decided by unanimous agreement that the mistakes he had highlighted should be examined using the astronomical instruments of the observatory. Verbiest and Wu were to undertake the observations simultaneously, recording in advance what they were observing and the method and instrument they were using. All this was to be accomplished in the presence of prominent Chinese astronomers and the presidents of the major ministries (twenty mandarins in all). A commission would report the results to the council, which would decide the next course of action.

The report offered to the emperor on February 22, 1669, was that "all Verbiest had predicted before, by calculation, answered exactly to the Heavens, and all Wu Mingxuan had proposed to observe, on the basis of his calendar, completely deviated from the Heavens." Nevertheless, the emperor, seemingly intent on embarrassing still more the enemies who had ruled during his childhood, declared himself unsatisfied with this report. Hadn't the same commission a mere four years earlier declared Schall's astronomy to be inaccurate? The commission then produced a second report that spelled out in clearer detail the science behind Verbiest's and Schall's work and why it was correct.

The emperor received this report at a seriously divided meeting of the council. This time, the two regents, Oboi and Ebilun, were in attendance. With the convincing second report to back him, the emperor, after sitting through a lengthy debate, took things in hand and ended the matter by ordering that Wu Mingxuan be brought in with his hands tied behind his back. While the two regents and their supporters grumbled, Wu was led off to the Ministry of Criminal Affairs to await his punishment. Yang Guangxian was spared severe punishment. In a confusing decree, the emperor declared that leniency was in order because Yang's "former accusation of Adam Schall was true"—an announcement probably made for unrevealed diplomatic reasons. Verbiest and his associates must have shuddered at that statement, but their own ordeal was over. In spite of the decree's odd assertion that Yang's accusation had not been in error, Schall, posthumously, was exonerated of all charges. Verbiest finally could write, "By demonstrating before those two rascals the mistakes of their calendar, I imposed on them the silence in such a way that they had nothing to answer."

Verbiest had triumphed, and so had the emperor. Before the year was out, Oboi, who by then had overcome Ebilun and taken complete control of the regency, was arrested. Young wrestlers leapt from a hiding place behind the throne in the emperor's audience room and overpowered him.

The Kangxi Emperor was, indeed, a dangerous man.

The calendar of the Chinese empire and the entire restoration of astronomy was entrusted to Verbiest. For him and his associates, this meant that meaningful scientific work could be resumed, although opposition to them would never completely disappear. The emperor honored Verbiest with several exalted titles, insisting on ignoring the fact that the Jesuit order forbade its members to take any official government appointment.

Now it was necessary to get rid of the thirteenth month. It would be no easy matter for any country—especially one as large and with as enormous a bureaucracy as China—to suddenly delete an entire month from the calendar year. Documents and dated contracts all had to be corrected and re-signed. Even more serious was the loss of face for China. The Minister of Rites asked Verbiest whether there might be some other way to rectify the problem, but the Jesuit astronomer prevailed with the argument that the Chinese calendar would be "contradictory to the Heavens" if the error were not corrected. The minister replied, "So it shall be, this month must necessarily be cancelled."

"It is hard to say," wrote Verbiest, "what novelties and changes this one intervention has brought with it, in so many matters!" The contests were over. The ceremonies began. One of the "novelties" to Western eyes was, in Verbiest's words, "the great pomp and solemnity" with which the Astronomical Bureau delivered the new calendars to the Imperial Palace: Early on the appointed morning, all the mandarins, in full ceremonial dress, gathered in the palace courtyards. Drums and trumpets sounded, and all the folding doors of the imperial gates opened simultaneously, admitting the mandarins of the Astronomical Bureau, who entered in a single, long row, in court robes and hats and the insignia of their ranks. A pavilion shaped as a large square tower with yellow silk curtains followed them, carried by forty bearers. Inside were the generously sized calendar volumes, bound in yellow silk, for the emperor, the empress, and the emperor's two consorts. Ten or twelve red-curtained, smaller pavilions followed with calendars—bound in red silk, interwoven with silver dragons—for the royal princes and other close relatives of the emperor. Next came tables draped with red carpets, holding volumes wrapped in yellow covers for all the other prominent people, such as those in charge of the various ministries. At a ceremonial pace, through doors at other times forbidden to all but the emperor himself, the procession moved along a

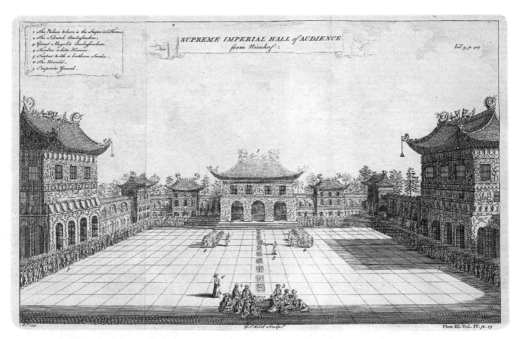

A Dutch engraving of the imperial audience hall and courtyard at the Forbidden Palace, Beijing, 1745.

path known as the Imperial Way, silk curtains billowing in the breeze, until the pavilions were at last set down in a prescribed arrangement. The mandarins of the Astronomical Bureau then removed the calendars from the first pavilion, placed them on tables covered with yellow silk, brought the tables to the gate of the Inner Palace, and presented the volumes to the household managers with three genuflections and three prostrations (a movement known as the *kowtow*). Returning to the courtyard, the mandarins of the Bureau then distributed the calendars from the lesser pavilions and tables to the intended recipients with more genuflections. At the conclusion of these rites, all returned to their places, according to rank. At a signal from the herald, they fell to their knees three times, three times stood up again, and at each genuflection bowed their heads three times to the ground. With that final reverent gesture, the ceremony ended and everyone returned home.

On the same day, similar ceremonies took place in the provincial capitals. Finally, the common people could purchase their own calendars. According to Verbiest, every household did, no matter how poor. Verbeist didn't say where he had been in the crowd that witnessed the delivery of his calendars. The emperor at one time offered to make him a mandarin "of the second rank," which was a high honor indeed, but that happened well after Verbiest first witnessed this ceremony.

The precocious teenager, now wielding full power as emperor, was developing into a man with exceptional physical strength and military skills, pouring seemingly inexhaustible energy into the daily administrative duties of his empire. He demanded no less exertion and devotion to duty from his chief astronomer. With the calendar delivered, Verbiest turned his attention to matters having to do not only with astronomy, mathematics, and mechanics, but also with the bureaucracy that dealt with astronomy and the calendar. Once or twice a month he produced astronomical reports for the palace in both Chinese and Manchu. The prediction of eclipses was an ongoing task that required his expertise and that of those trained to assist him. Such predictions differed for the seventeen districts of China, and all had to be calculated six months in advance so that the proper rites could be planned. These rites, with hardly less ceremony than the calendar presentation, included the beating of drums and cymbals and the playing of other instruments to create a din that echoed through the city. According to ancient tradition, the noise supported the sun and moon in their "time of trouble," although Verbiest pointed out that Chinese astronomers understood eclipses very well and knew that this pandemonium could not in the slightest affect the destiny of the sun or moon.

The Kangxi Emperor grew up to be an exceptionally effective ruler, defeating or immobilizing enemies and potential enemies, and securing Qing dominion over all of China. By both military and diplomatic maneuvers he won back his dynasty's Manchu homeland, and Manchuria became part of his empire. In 1696, after Verbiest's death, he personally led an extremely dangerous military expedition across the hostile, scorching Gobi Desert to add Outer Mongolia. Cannons designed by Verbiest played an essential role in these campaigns.

The emperor's skill in management within his empire was no less effective. His love of knowledge in many fields, including a passion for European art and painting, never waned. Verbiest was a trusted friend and mentor who knew that this powerful man was still something of a child at heart, so he presented him with a truly remarkable toy.

In a memoir of his scientific and technological work during twenty-five years in China, Verbiest mentioned an invention that, though he evidently considered it of no great significance, seems especially impressive to us today: the first automobile. It was, in fact, this little steam-driven device that first called Verbiest to my personal attention. Its design came to light in Europe in the late 1680s, when a Bavarian publisher, Johann Kaspar Bencard, published a beautiful edition of Verbiest's *Astronomia Europaea*.

Verbiest reported that he was experimenting with an "aeolipyle," a device dating from ancient times in the West. Basically, an aeolipyle is a closed vessel in which water is heated until it creates steam that escapes through small nozzles. If the aeolipyle is placed on a pivot, two nozzles faced in opposite directions can turn it. Verbiest's aeolipyle was cylindrical, heated by burning coal beneath it, and fixed in place so that it could not rotate. He used only one nozzle for steam to escape, and he positioned in front of the aeolipyle a horizontal wheel one foot in diameter, with small, wing-like double bars on its outer rim. The steam expelled from the nozzle of the aeolipyle blew against these wings at an angle, causing the wheel to turn, something like the way falling water causes a waterwheel to turn.

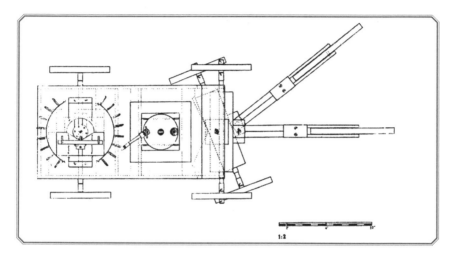

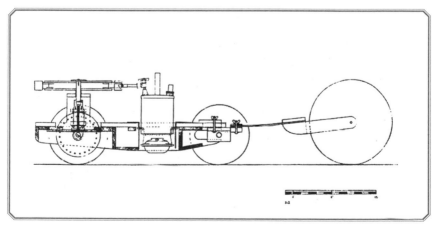

Line drawings illustrating Verbiest's "automotive machine," drawn by Ditlev Scheel, 1994.

Verbiest constructed his two-foot-long, four-wheeled vehicle out of light wood. The wheels were probably the delicate, spoked wheels popular in China at the time. Verbiest fixed a vertical wheel (which didn't touch the ground) to the front axle of his little vehicle in order to turn the front wheels. The teeth of that vertical wheel engaged the teeth of a horizontal wheel above it, which was fixed below the "winged" horizontal wheel. When steam from the aeolipyle caused the upper horizontal wheel to turn, the second horizontal wheel, having teeth rather than wings, moved with it and engaged the teeth of the vertical wheel fixed to the axle. Verbiest recorded that the vehicle could go for an hour or more at a considerable speed as long as the coal burned and the aeolipyle continued to produce steam.

In order to control its direction, Verbiest modified the vehicle so that the rear axle could pivot and attached a tiller to the center of that axle. At the end of the tiller he attached a large, freely moving wheel. By turning the tiller to the right or left and fixing the position with a screw, he could cause the vehicle to move in larger or smaller circles, depending on the angle of the tiller and rear axle. Now he could adjust his "automotive machine" to run within the dimensions of a room or courtyard. Verbiest then added a whimsical refinement: a second nozzle on the aeolipyle, with a whistle, which produced the song of a nightingale. He also fashioned a set of bells into a miniature carillon that played melodies.

Verbiest's inventiveness was not yet exhausted. Employing a similar "method of movement," he caused a little paper boat to sail in circles with bellying sails, and he hid the mechanism so that it seemed the faint hiss of the steam was the sound of the water rushing around the boat. One cannot doubt the delight of the emperor (and, in the case of the boat, the emperor's brother) when Verbiest presented and demonstrated these marvelous toys. However, Verbiest's automobile was too small to carry a human passenger, and a twentieth-century writer's report that Verbiest rode his automotive machine through the streets of Beijing is definitely a flight of imagination.

Such mechanical marvels—and the Kangxi Emperor's seemingly insatiable curiosity about such things—sealed even further the relationship between the two men as the years passed. As Verbiest described in the language of allegory,

After Astronomy, marching like a venerable queen between the
Mathematical Sciences and rising above all of them, had made
her entry among the Chinese and had ever since been received
by the Emperor with such an amiable face, all the Mathematical
Sciences also gradually entered the imperial Court as her most
beautiful companions . . . as if they were gold and precious stones,
to find more favour in the eyes of such a great majesty.

Verbiest's output in the way of astronomical studies, inventions, and magnificent astronomical instruments was prodigious. He sent some of his books, in Latin, to Europe, but most of his writings were only in Chinese and, in his own time, never went beyond the borders of China. In 1669, under his supervision, the old instruments of the Yuan and Ming eras were taken down from their platform on Beijing's eastern wall and replaced by at least nine new instruments, six of them exquisitely made, elaborately decorated, and designed by Verbiest himself in the style of Tycho Brahe.

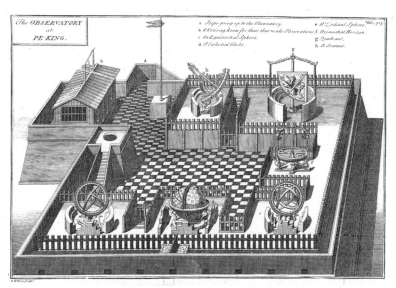

An engraving of the platform observatory in Beijing (today known as the Beijing
Ancient Observatory) from Du Halde's The General History of China. *The astronomical*
instruments depicted are (clockwise from top left) a sextant, a quadrant, an azimuthal
horizon, an equinoctial sphere, a celestial globe, and a zodiacal sphere.

However, Verbiest's difficulties were not over. The emperor's insistence on showering him with exalted titles conflicted with the Jesuit vow not to aspire to high dignities outside the Jesuit order. At one point even his faithful colleagues Buglio and de Magalhães became openly hostile. They were no doubt inspired in part by justifiable envy: The emperor was giving Verbiest credit for all the best work of the Jesuit astronomers with complete disregard for the substantial and essential contributions of his two associates. But their opposition was also motivated by a sincere and reasonable fear that a foreigner's holding such exalted positions would bring reprisals within the slippery, xenophobic imperial court.

But there was another side to the argument, with which an overwhelming majority of Jesuits and other Christian leaders all over China agreed wholeheartedly: To refuse or renounce honors conferred by the Kangxi Emperor would have been taken as a gross insult, ending Verbiest's valuable relationship with him and probably resonating far beyond Beijing and the Jesuit order to jeopardize the success of the entire Christian enterprise in China. Eventually the Vatican declared Verbiest innocent of any wrongdoing and instructed him to continue his official dignities "without further molestation." Researchers who have delved most extensively into the correspondence and reports having to do with this difficulty, which lasted many years, have been awed to find that "in all the sources we possess, Verbiest's pen does not produce even one word of bitterness, anger, or recrimination." Clearly, he was an extraordinarily focused, self-assured, patient man.

It was on the occasion of an inspection of a set of cannons Verbiest had designed and cast that the Kangxi Emperor turned aside and asked to see Verbiest's church and residence. After a two-hour visit, the emperor "by his own hand" created two large Chinese characters and sealed them with the imperial stamp as a gift to the church, where "they render[ed] us great prestige and respect."

Verbiest worked at the Beijing Observatory until his death at age sixty-five on January 28, 1688. In his final communication with the emperor, he wrote,

Sire, I die content, since I have used almost all the moments of my life in the service of Your Majesty. But I very humbly beg Your Majesty to remember after my death that in all I have done, I have had no other view than to procure, in the person of the grandest king of the Orient, a protector for the holiest religion in the universe.

His dying hope temporarily became a reality in the Chinese Edict of Toleration in 1692, which officially permitted the preaching of Christianity in China. Sadly for his legacy, controversies involving the papacy and the Jesuits over such matters as whether a Chinese who became a Jesuit could administer the Catholic sacraments, whether it was right to have Christianity accepted as one of four religions tolerated in China, and disputes among missionaries of different orders, would plague the Christian church in China for many years.

With the eventual failure of the Jesuit Chinese mission, and the triumph of Copernican over Tychonic astronomy in the West and eventually in China, Verbiest's earnest endeavors essentially came to little. His beautiful instruments still stand in Beijing, but they are artifacts, and his name has been almost entirely forgotten. The Kangxi Emperor's has not.

A 2009 photograph of the Beijing Ancient Observatory. Instruments from left to right are the azimuth theodolite (1715), altazimuth (1673), ecliptic armillary sphere (1673), celestial globe (1673), and "new" armillary sphere (1744).

TWO

Benjamin Thompson, Count Rumford

FARM LAD, SPY, ARISTOCRAT, RASCAL

(1753–1814)

I N THE AUTUMN OF 1775, a few months after the Battle of Bunker Hill in the American War of Independence, New Englander Benjamin Thompson settled his affairs, sold his land, cleared his debts, collected debts owed him, and made a hasty departure from the town of Woburn, Massachusetts. His stepbrother Josiah Pierce took him to Narragansett Bay, allegedly to embark for the West Indies. Instead, he boarded a British ship at Newport and sailed to Boston. It was an unusual, roundabout way of reaching a city only eleven miles south overland from Woburn, but Thompson had his reasons. He had been spying for the British; and though a Committee of Correspondence had failed to convict him, suspicion about his activities had grown uncomfortably strong among his neighbors. He had been cautious, writing dispatches in invisible ink he'd invented from nutgalls on the bark of oak trees. (He had devised a "cover" for collecting them by pretending to be plagued by recurrent diarrhea, for which they were a known cure.) Nevertheless, after one narrow escape from tarring and feathering—or worse—it was time to head for safety.

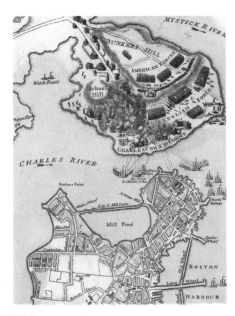

OPPOSITE: *An engraved portrait of Benjamin Thompson, Count Rumford, c. eighteenth century.* RIGHT: *An inset of a map of Boston entitled* The Seat of War in New England . . . with the Attack on Bunker Hill, *printed in London in 1775.*

The British evacuated Boston the following March, and Thompson sailed to England, carrying dispatches to Lord George Germain, secretary of state for the colonies. Benjamin Thompson may be almost entirely forgotten in the history of science, but he was not a man to let himself be overlooked during his lifetime. He was an assiduous self-publicist, and he made certain that the documents he carried to London included a complimentary letter recommending him to Germain from British general William Howe in Boston.

The spy-turned-envoy had already risen far from the Woburn farmhouse where he had been born in March 1753. His formal schooling had been hit-and-miss, but he had done a good job of educating himself by badgering knowledgeable acquaintances with questions and winning the friendship of a "very respectable minister" who taught him algebra, geometry, astronomy, and some higher mathematics. Somehow he also acquired exceptional abilities with language and early on displayed a gift for understanding and inventing small, practical mechanical devices (not unlike another eighteenth-century Benjamin: the polymath Benjamin Franklin). He learned to sketch and caricature people and to play the violin, beginning a lifelong passion for music.

Benjamin Thompson's childhood home in Woburn, Massachusetts, shown in a photograph from 1874.

Samuel Curwen, next-door neighbor to the importer where Benjamin was apprenticed at age thirteen in Salem, Massachusetts, would later comment that the boy displayed a "fondness for experimental philosophy [modern-day science]." That was a gentle way of putting it. While manufacturing fireworks for a community celebration, Benjamin had set off an explosion that could easily have cost him his eyesight, if not his life. But Curwen liked Benjamin: "This young man . . . ever appeared active, good natured and sensible." At age fifteen Benjamin calculated an eclipse, correct to within four seconds. Many years later, in his memoirs, he recalled reading the Dutch physician and chemist Herman Boerhaave's *Treatise on Fire* when he was seventeen, as the beginning of his lifelong interest in the nature of heat: "Subsequently, indeed, I was often prevented by other matters from devoting my attention to [experiments with heat], but whenever I could snatch a moment I returned to it anew, and always with increased interest."

During his apprenticeship to the merchant John Appleton, whose family were members of Salem's social and intellectual elite, Benjamin thought it in his interest to cast off some of his farm-boy persona and emulate the manners and accomplishments of a well-educated, well-off family. When apprenticed later in Boston, he made time to acquire other skills that would make it easier to slip into an aristocratic role. A Boston "private school" taught fencing and French, as it advertised, "in a most concise manner and on reasonable terms." He enrolled for both.

By age nineteen, Benjamin had picked up some medical knowledge as an apprentice to a doctor, attended a few science lectures at Harvard College, and evidently made a sufficiently educated impression to be offered a position as schoolmaster in Bradford, Massachusetts. There, taught privately by Rev. Samuel Williams, who later became Hollis Professor of Mathematics and Natural Philosophy at Harvard College, he first studied science seriously. But Benjamin's stay in Bradford wouldn't last long: The Reverend Timothy Walker, who had known him in Woburn, invited Benjamin to help found a school in Concord, New Hampshire. With that invitation, Benjamin's destiny took a significant turn.

Elementa Chemiae *by Herman Boerhaave, published in Leiden, Holland, in 1732. It includes his "Treatise on Fire," which inspired young Benjamin.*

At Concord—formerly called Rumford—Benjamin Thompson's meteoric rise began. From that time onward there would be a recurrent theme running through his story: remarkable skill and ease in cultivating influential friends, including wealthy women, and no compunctions about using them to his own advantage. He was extremely personable. His closest friend as a teenager and for the rest of his life, Loammi Baldwin, described him as "of a fine manly make and figure, nearly six feet in height, of handsome features, bright-blue eyes, and dark auburn hair [and] the manners and polish of a gentleman, with fascinating ways, and an ability to make himself agreeable." He also had a driving ambition and innate ruthlessness—qualities that would help him rise to the top of almost any social or professional situation, capitalize on any opportunity to further his own interests, and usually escape the worst consequences of his mistakes. Those same attributes made him a not particularly likable man, quick to lose friends, often hated, and destined to watch moments of well-deserved triumph disintegrate largely because of his arrogance and certainty that he was always right—which he usually was.

Concord, New Hampshire, had far less to offer than Boston in the way of intellectual stimulation, but it did have something else. Sarah Rolfe, daughter of his sponsor Reverend Walker, was a widow with one son. Her husband had been the richest landowner in Concord, a leader in New Hampshire colonial society, and a colonel in British governor John Wentworth's militia.

Whatever genuine affection existed between Sarah and Benjamin, she, in her mid-thirties, was getting old, and Benjamin longed to escape a semi-impoverished future. Their engagement was a godsend for both.

Thompson wasn't yet twenty when the marriage made him proprietor of two thirds of the land in Concord, and he quickly abandoned teaching to assume the role of a gentleman. Sarah and he journeyed to Boston to a tailor and hairdresser to fit him out in a manner appropriate to his new rank. Returning in a fashionable two-wheeled chaise, they stopped by his mother's door in Woburn. The astonished woman scolded her son: "Why, Ben, how could you go and lay out all your winter's earnings in finery?" This finery was part of an extensive and expensive new wardrobe.

As a major in Wentworth's regiment—a promotion above older and more experienced men that didn't make him popular—Thompson proved himself an able loyalist. He organized a technique for rounding up deserters from the British army (and saw to it that Wentworth mentioned this achievement in dispatches to the Earl of Dartmouth in London). However, in a New England where revolution was brewing, a commission in the British Army, adopting an aristocratic demeanor, socializing among British aristocrats, and rounding up deserters who preferred to join the American side made Thompson the object of so much suspicion and

unpleasantness that he found himself facing Concord's "Committee of Correspondence," charged with being a "Rebel to the State." There wasn't enough evidence to convict him, but mobs intent on tarring and feathering ignored formalities.

A nineteenth-century engraving of John Wentworth, British governor of New Hampshire (1767–1775), after a portrait by John Singleton Copley. Thompson served as a major in Wentworth's regiment.

Thompson fled to Boston on a horse lent by his brother-in-law Timothy a few hours before the mob arrived, with twenty dollars in his pocket. Timothy, left with Thompson's wife and two-month-old daughter, barely managed to persuade the angry citizens to spare the house. Thompson wrote an apologetic letter to his father-in-law, who disapproved of his Tory friends and leanings. He included the pledge, "This you may rely & depend on, that I never did, nor . . . ever will, do any action, that may have the most distant tendency to injure the true interest of this my native Country." Whether that "true interest" was remaining a British colony or achieving independence, he left unsaid.

In Boston, it wasn't long before Thompson caught the attention of British General Thomas Gage, who was recruiting spies. Not only did Thompson have military experience, but also he had grown up in New England, so he knew the area and his accent was not clearly British. He was quick-witted, clever, personable, and eager. In short, he was the ideal candidate. He took lodgings around the corner from Isaiah Thomas, printer and publisher of a revolutionary newspaper. Revolutionaries including Paul Revere often met in Thomas's house and shop. Thompson made his approach through Thomas's wife, Mary, and began an affair with her that provided ample opportunity to learn what was discussed at the meetings. He was clever, even able to return to Woburn for a while before making his hasty, final roundabout escape to Boston and then on to England. Again he wrote to his father-in-law, this time completely abandoning his wife and daughter to the old man's care. Sarah had served her purpose. So had Mary Thomas.

In London, Thompson began a new life on a higher stratum of society than he had enjoyed in New England, even after his marriage. General Howe's letter, coupled with Thompson's own charm, skill, and his reliable and rare knowledge of what was happening in the Massachusetts Bay Colony in America, served him admirably. Germain appointed him his private secretary, and within three years he was an undersecretary of state. Not that Thompson made an unfailingly good impression among his

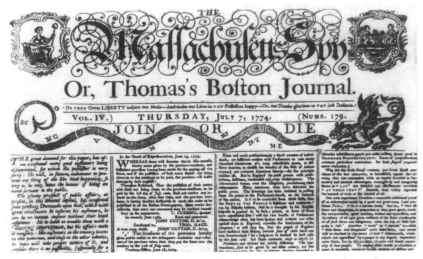

The front page of the July 7, 1774, edition of The Massachusetts Spy, or, Thomas's Boston Journal, *a revolutionary newspaper published by Isaiah Thomas. As a British spy, Thompson began an affair with Thomas's wife, Mary, in order to learn what was discussed at revolutionary meetings held at Thomas's shop.*

English acquaintances. Germain was not well-liked, and Thompson suffered by association. It didn't help that he built a rather shady reputation among the Tory American expatriate community in London as the man through whom favors from British officials could be sought. His price was steep and results were not guaranteed. One Dr. John Jeffries recorded in his diary that he was amenable to his wife spending night after night for three months with "Mr. Thompson of Pall Mall." The favors Jeffries hoped for never materialized.

Thompson's first scientific experiments in England at Germain's estate in the summer of 1778 tested the force of gunpowder in a new way—a development of major value to the military and in line with his continuing interest in heat. Thompson submitted a paper to the Royal Society of London (then known as the Royal Society for the Improvement of Natural Knowledge) under the lengthy title "New Experiments upon Gun-powder, with occasional Observations and practical Inferences; to which are added, an Account of a new Method of determining the Velocities of all Kinds of

Military Projectiles, and the Description of a very accurate Eprouvette for Gun-powder." In 1779 the Royal Society elected him a fellow, recording that this largely self-taught twenty-seven-year-old was "a gentleman well versed in natural knowledge and many branches of polite learning." He would go on to publish more than seventy papers in his lifetime.

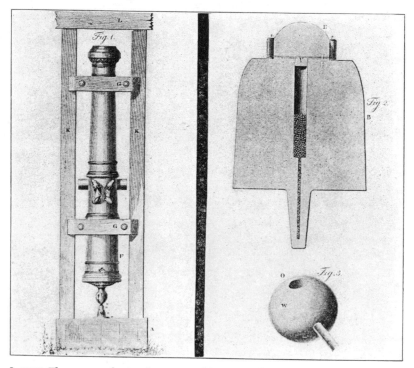

In 1778, Thompson submitted a paper on his gunpowder experiments to the Royal Society, which elected him a fellow in 1779. Here, a drawing of one of the experiments.

Thompson's espionage skills were not growing rusty. Germain, hoping to wrest control of the British Royal Navy from John Montagu, fourth Earl of Sandwich and First Lord of the Admiralty, sent Thompson to sea on a ship in the Channel Fleet, ostensibly to continue his study of gunfire but actually to report on the state of the Royal Navy. Another activity that was rather underhanded but very profitable was serving as the administrator for the provision of clothing to the British forces in the colonies. Silk was used extensively in uniforms, and Thompson knew that silk could be expected to gain weight in a voyage across the ocean.* Through buying silk by weight before a transatlantic journey and selling it at its new, greater weight when it arrived, Thompson made a hefty profit, more than replacing the fortune he had abandoned in New England. His old neighbor Curwen would comment that the former humble apprentice now had an income "near seven thousand a year—a sum infinitely beyond his most sanguine expectations."

By 1780, the war in America had become a disaster for Britain. In London, unpopular figures like Germain and Thompson maneuvered to save their positions and reputations. Thompson found himself in serious danger when Frenchman François Henri de la Motte was caught with British naval plans. La Motte was drawn and quartered in a public spectacle. His source was never discovered, but many suspected Thompson. He had become a persona non grata in London.

It was once again time to wrap up his affairs and look around for a fresh opportunity. He found it back across the Atlantic, where Britain's lost cause still offered a chance to prove his loyalty and ability. In New York he took command of his own paid army—the King's American Dragoons—and arranged, at considerable expense, for the official recognition of his dragoons to take place during a visit in August 1782 from King George III's sixteen-year-old son, Prince William Henry. The prince reviewed the Dragoons and presented the colors "with his own hand" to Thompson, who made sure that this news item reached London newspapers.

Thompson's glowing dispatches sent from America were mostly fiction (unlikely to be checked in the confused aftermath of the war). He did, however, manage to come up with two ingenious, useful inventions: a gun carriage that could be disassembled, carried on the backs of three horses, and reassembled in approximately seventy-five seconds; and a device made of cork that allowed a horse to swim with a cannon on its back. One of Thompson's reports boasted that there had been very few complaints from the local people where the dragoons were stationed. A secret Agent "W"

* Under differing circumstances, silk either loses or gains weight during a voyage over water, but in the humid hold of a late-eighteenth-century sailing ship, the silk was much more likely to absorb moisture and gain weight than to lose it.

reporting to the American side told another story, and so did the long memory of the inhabitants of Huntington, Long Island. Thompson became infamous for tearing down their church and forcing local men to build his fortifications with the salvaged material. He ordered 142 apple trees cut down and requisitioned all the chestnut fence rails. When his dragoons finally marched away, he ordered all this wood burned so that the towns-people would have neither fence rails nor firewood.

This political cartoon by noted British satirist James Gillray, published in February 1780, is entitled "The State Tinkers." It skewers Lords North, Germain, and Sandwich, as well as George III, and their handling of the war.

Thompson's offenses did not stop there. He built barracks on the burial ground and pitched his tent on the grave of the beloved vicar who had built the church. Valuable old books from the present, aging pastor's library were strewn on the stable floor to be trampled. His men used gravestones as baking ovens, making a joke of delivering bread to the townspeople with names of their dead relatives indented in the crusts. The locals regarded him as "the devil incarnate." Long Island historian Nathaniel Prime, writing in 1845 within living memory of these events, lamented, "His acts in this place have given him an immortality which all his military exploits, his philosophical disquisitions, and scientific discoveries will never secure to him among the descendants of this outraged community." Although Thompson was certainly not always an admirable or honorable man, stories like this were rarely associated with him. The American Revolution had been, in his eyes, an egregious betrayal of his "native country." Perhaps this was his act of relatively small-scale vindication, an ugly revenge on a civilian population—but without taking lives.

When the preliminary articles of peace ending the American Revolutionary War were signed in November 1782, Thomson prepared to take advantage of the reputation he had been rebuilding in England to secure a high-level and financially secure future. When he returned to England in April 1783, thirty-year-old Thompson planned to return to his scientific studies concerning the nature of heat, but he learned that "Wilkin, in Sweden, had already carried out exactly what I had proposed to myself. . . . I laid aside, as useless, the apparatus which I had designed for my own investigations."

A second plan, to offer his military skills to Austria in her war with Turkey, also fizzled when that conflict began to wind down. By that time, Thompson had instead requested permission to serve at the court of Karl Theodor, elector of Bavaria. England needed an effective representative there, and Thompson was quick to point out that the more exalted a personage they sent, the more likely he was to have access to the highest circles. This argument made sense, so King George III granted him

Portrait of Charles Theodore (Karl), Elector of Bavaria (1724–1799), by eighteenth-century Prussian painter Anna Dorothea Therbusch, 1763.

knighthood. By the summer of 1785, thirty-two-year-old Sir Benjamin Thompson was in Munich with the title of chamberlain. The Royal Society had a promise from him to share anything he found of scientific interest. All seemed to be going very well for Thompson, even though his best link with high levels of power in England ended when Germain died that August.

Munich loved Thompson. Often seen mounted on one of the thorough-bred horses he had brought from England, he made a splendid appearance—an imposing figure with a handsome face and dignified, courteous manners "not only to the great, but equally to subordinates and inferiors." He learned to speak German and French and began making major changes in military organization, discipline, and tactics, setting up improved breeding schemes for horses, bettering conditions and education for military families, and lending his inventive skills to the development of arms and ordnance. All of this endeared him to Karl Theodor, but it increasingly created enemies among the military establishment with vested interests in the status quo.

It was in a remarkable turnabout that a new aspect of Thompson's character and skills appeared. Bavaria was a well-known slough of poverty and crime. Beggars (often also, out of necessity, thieves) haunted the towns and villages. On New Year's Day, "from time immemorial considered in Bavaria a day peculiarly set apart for giving alms, and the beggars never failing to be all out upon that occasion," Thompson rounded them all up. He had been pondering the old belief that people had to be virtuous in order to be happy and decided to turn it around. He would try to make the wretched beggars of Bavaria virtuous by making them happy *first*. Accordingly, the former beggars found themselves set to work in factories producing army uniforms, their comfort and cleanliness a priority, with good and plentiful food, fair pay, and education—optional for adults, required for the children. The scheme worked amazingly well, and many former beggars celebrated Thompson later for what he had done for them. He also arranged that "widows and unmarried ladies of very small fortunes" be given raw materials to do piecework in their homes, receiving pay and meals. A handsome building was soon erected where bastard children of the nobility could be born and subsequently educated and prepared for entrance into the military academy.

Though his innovative social work might seem to indicate otherwise, Thompson's attitude toward "men in the mass" had not changed. He was by no means a "leveler"—an idea he ridiculed in America. He

> thought it was not wise or good to entrust to men in the mass the care of their own well being. The right, which seems so natural to them, of judging whether they are wisely governed, appeared to him to be a fictitious fancy born of false notions of enlightenment. . . . Government should be left in absolute control of intelligent men alone, the will of a few sound heads.

It goes without saying that Thompson considered himself, with some justification, one of the "intelligent men," one of the "sound heads."

Thompson's workhouse opened up a new line of scientific inquiry. The building needed to be well lit with artificial light. He reported on the project in a paper, "An Account of a Method of Measuring the Comparative Intensities of the Light Emitted by Luminous Bodies," and was responsible for coining the term *candle power*—used into the twentieth century as the international unit of measurement for illumination. Related to his favorite subject, Thompson was also designing instruments and experiments to study the propagation of heat through various substances. In 1786, the Royal Society rejected his long-winded paper reporting that dry air is a better insulator than moist air, but a 1792 paper reporting his discovery that it is the amount of air and how it is trapped in cloth that determines the material's insulating properties won him the Royal Society's highest honor: the Copley Medal. The uniforms produced in his workhouse were modified to take advantage of this new understanding.

Heat and light were not Thompson's sole concerns. In a debate about the "green matter" in standing water, he argued ahead of his time that it was of an "animal nature, the assemblage of an infinite number of very small, active, oval-formed animalcules." Although he didn't follow up on it, the idea was not wasted. He entertained onlookers with experiments in the palace gardens.

Karl Theodor's court was casual about who was whose mistress. The Countess Baumgarten—an amply built, buxom beauty—was Thompson's *and* the elector's mistress. In 1789 she bore Thompson a daughter, Sophy. Another of his mistresses was Countess Baumgarten's sister, the Countess Nogarola, She was no beauty but intelligent, with intellectual interests paralleling his own, and they remained friends for many years. She translated some of his work into Italian and helped him write in French and German.

By the time Thompson had served the Bavarian court for nearly eleven years, major innovations he had brought about in their bookkeeping and funding amounted to little less than a reorganization of the Bavarian economy. To counter the festering enmity his changes were inspiring, he gave the city of Munich an enormous, public "English garden" and celebrated the beginning of the project with a festival and free grand ball for the entire city. His soldiers, Thompson in the lead, marched in carrying garden tools instead of weapons. The elector was delighted. More honors—and still stronger efforts by Thompson's opponents to undermine him—followed. This time, he saw to it that some lost their positions and their fortunes. In 1790, Theodor named him a count of the Holy Roman Empire. He chose the title "Count Rumford," using the older name of his New Hampshire town, Concord. At the same time, the number of his enemies was growing, and they were increasingly powerful.

By 1792, Elector Theodor was elderly, his court an unpleasant hotbed of intrigue, and it looked likely that Austria would take over Bavaria.

A portrait of Count Rumford engraved by J. P. P. Rauschmayr, 1792. Thompson took the name "Rumford" from an earlier name for Concord, New Hampshire, where he began his rise to prominence.

Thompson decamped again, this time on temporary "leave" to Italy, where he resumed his study of heat in a surprisingly down-to-earth, practical way. He turned his attention to devising more effective ways of using fuels to warm rooms and cook food—the layout of kitchens; how stoves should be built; how coal grates, stovetops, and chimneys could be improved. Domestic science classes still study some of his precepts, and his idea to include a round hole in the end of the handle of a pot, by which it can be hung when not in use, has certainly become a standard. A Rumford kitchen had the burners recessed in the cooktop with the top rims and lids of the pots flush with the surface of the stove, so that the flames could travel up the sides of the pots, heating them as well as the bottoms. He arranged these burners in a semicircle around the cook, each within easy reach without stretching over other burners.

When Thompson had been absent from Munich for a year, his excuse for leaving—ill health—was growing weak, and he began the return journey. However, it soon became clear that it was not advisable for him to remain in Bavaria, for Theodor, though still elector, was completely under the thumb of Thompson's enemies. In March 1795, he went back to London.

The most successful project of Thompson's life began with a request from Lady Palmerston. He'd had an affair with her in Florence, and in London they were seeing each other again. She wanted to redecorate, but what was the use? Everything would soon be sooty again! Couldn't her lover use his scientific prowess to improve her fireplace? Thompson's solution caught on not only in England, but also in Scotland, Europe, and America.

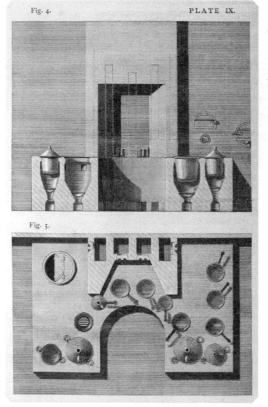

Fig. 4.

PLATE IX.

Fig. 5.

LEFT: *Diagrams showing front and overhead views of a Rumford kitchen with an open chimney fireplace, from Thompson's essay "On the Construction of Kitchen Fireplaces and Kitchen Utensils," 1800.* OPPOSITE, LEFT: *Front and overhead diagrams of a Rumford-altered chimney, from Thompson's essay, "Of Chimney Fireplaces," 1796.* OPPOSITE, RIGHT: *Diagrams showing a side section of a chimney before (top) and after (bottom) being altered with Thompson's modifications.*

In 1796, when he published "Of Chimney Fireplaces," no fewer than two hundred fifty fireplaces were converted in London alone. A "Rumford" became a must for almost anyone who could afford one. Thomas Jefferson had Rumford fireplaces at Monticello.

On first encounter, Thompson's design improvements didn't appear likely to be improvements at all. It was assumed that the larger the fireplace, the more wood it could burn, and the more heat it would provide. Rumford insisted it is not the size of the fire or fireplace, but rather the amount of heat radiated into the room that matters. He replaced the huge box-like opening with a small, shallow fireplace, the two side walls set at angles converging toward the back. A fire in this space arrangement required much less fuel, and much more heat was reflected into the room.

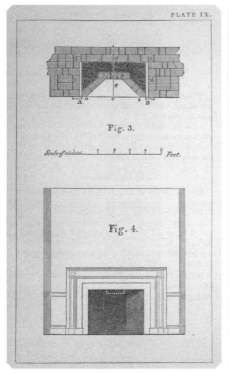

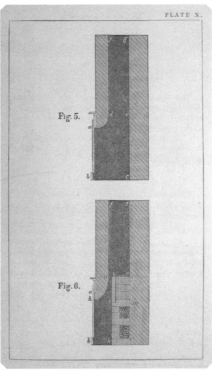

Thompson's design was also much more effective in dealing with the smoke, as Lady Palmerston had hoped. It seemed that the larger the chimney, the more smoke it ought to carry up and out, but air turbulence in big chimneys caused an unpleasant amount of smoke to be expelled into the room. Big chimneys also carried away heated air. Thompson's chimney had a small throat—four inches (ten cm) across or less—curving smoothly up behind the mantelpiece. The size of his fireplaces and chimney openings must have seemed ludicrous to those who lived in houses with large, chilly rooms, but Rumford fireplaces work amazingly well. Conversions were not difficult but had to be done correctly and with care in order to achieve the desired effect. Rumford wrote two papers describing in detail his improved fireplace designs in 1796 and 1798, but potential customers, concerned that the conversion be done right, still sent inquiries and invitations for him to send crews of bricklayers to fine homes all over the British Isles. In Jane Austen's

1818 novel *Northanger Abbey*, seventeen-year-old Catherine Morland enters the house anticipating a rather spooky, ancient abbey but discovers that "the fireplace, where she had expected the ample width and ponderous carving of former times, was contracted to a Rumford." Rumford fireplaces had clearly entered popular culture by the nineteenth century. Two centuries later, they are still being manufactured in the United States, and you can spot them in room after room when touring many of the great English estate houses now preserved by the National Trust and English Heritage.

A related discovery made large fireplaces even less of a necessity. Thompson wrote to Lady Palmerston that his kitchen maids had stored meat overnight in the machine he had invented for drying potatoes, and in the morning found it perfectly cooked but "different both in taste and flavor from any I had ever tasted before." Meat cooked in an enclosed oven, very slowly and at a low heat, meant lower cost. To make the temperature constant and adjustable, he designed a roaster with blowpipes to regulate the flow of air. Rumford roasters, like his fireplaces, were soon in use all over Europe and in America. He wrote detailed instructions for building large ovens for such establishments as his workhouses. "I lately roasted one hundred pounds of veal," he recorded, for "something less than one penny farthing."

Each year on her birthday, Thompson had been sending his mother the interest on a $500 investment. In November 1795, he sent her a $5,000 U.S. government bond, with permission to use the interest or principal however she pleased. He also utilized his new wealth to bring his twenty-one-year-old American daughter Sarah (whom he called Sally) to London and quickly dispatched her to finishing school when her ignorance of fashionable etiquette embarrassed him. He donated funds to the Royal Society in London and to the American Academy of Arts and Sciences in Boston, where the interest on his $5,000 gift was to be awarded each year as a prize to the "author of the most important discovery or useful improvement" having to do with "heat or light." His prizes are still given.

When Thompson's official leave from the Bavarian court expired, Elector Theodor lured him back with promises to restore his former status.

Thompson took Sally with him, carefully briefed by Lady Palmerston on what would be expected of her at court. This time it was Thompson's military expertise that was required. The city of Munich was about to become a battlefield between the Austrian and French armies, and an evacuation was in progress. Art treasures and official documents had been loaded on barges and taken to Passau by way of the Isar and Danube rivers. The elector and his closest associates were fleeing to Dresden, and Theodor left his city in Thompson's hands. "It was settled that I should remain at Munich," Thompson wrote to Lady Palmerston,

> that I should have an eye to what should be going on—that I should correspond constantly and confidentially with the Elector—and that, in case the French should come to Munich, I should endeavor . . . to interest the feelings of their Generals in . . . the preservation of the Town and the Country.

Thompson was confident that he could keep both the Austrian and French armies from invading Munich, and all went well as the Austrian general, instead of invading the city, went around it to meet the French. But then the Bavarian general blundered disastrously by insulting him, and the Austrian general changed tack and entered the city instead. Thompson rapidly assumed command of the defense of Munich and organized the food supply and mass feeding for the people and the military during the siege. He invented a portable stove for use inside a tent or atop the city walls. Despite the enemies who disparaged Thompson, it is clear that throughout his years in Munich he served the city and Bavaria splendidly.

In 1798, Thompson was back to his science, investigating the way heat travels in liquids. Integral to his chimney design was his understanding of what we now call *convection currents*. They were previously understood to occur in air, but Thompson was the first to call attention to their role in heat conduction in liquids. It was another study, however, that won him a place—though his contribution is now almost forgotten— in the history of science.

One of Thompson's responsibilities in Bavaria was to oversee the foundry where cannons were made. "It frequently happens," he observed, "that in the ordinary affairs and occupations of life, opportunities present themselves of contemplating some of the most curious operations of Nature." His conclusions regarding these particular "curious operations" in the foundry would be the capstone of his lifelong interest in the nature of heat.

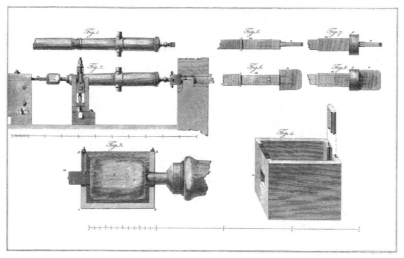

Figures illustrating Thompson's cannon-boring experiments from his "An Inquiry concerning the Source of the Heat Which is Excited by Friction," 1798.

The foundry cast each brass cannon as a solid brass cylinder. The hole through which cannon balls would be shot was then ground out, or bored, with an iron tool (iron being harder than brass), so that the brass cylinder became a tube. This was achieved by holding the iron tool steady in place while the brass cylinder was rotated by two circling horses. The cylinders rapidly became extremely hot, and the chips that came off them in the process were even hotter. "The more I meditated on these phenomena, the more they appeared to me to be curious and interesting," Rumford wrote. "A thorough investigation of them seemed even to bid fair to give a farther [*sic*] insight into the hidden nature of Heat." He submerged part of the cylinder in a watertight wooden box and filled the box with water to measure the heat produced. As the boring proceeded, the water grew hotter until, two hours and thirty minutes after the beginning of the experiment, Thompson enthusiastically reported, "It actually boiled!"

The cylinder didn't run out of heat as the process continued. The heat resident in it was apparently inexhaustible. We might ask why anyone would think it should run out, and the answer is that the prevalent theory to explain heat in Thompson's time was the "caloric" theory, which considered heat to be a substance, like a liquid, that was stored in a material such as the brass in the cylinder and that could be squeezed out by some type of work (e.g., the horses). Eventually, if the caloric theory were correct, there would be no more heat in the brass of the cylinder. But Thompson decided to reject that theory in favor of an older one in which heat is a form of motion. As he wrote, not claiming by any means to be the first to think of it, "I must confess that it has always been impossible for me to explain the results of such experiments except by taking refuge in the very old doctrine which rests on the supposition that heat is nothing but a vibratory motion taking place among the particles of the body." In so successfully challenging the caloric theory, Rumford made a leap into the next century's understanding of heat. There was not yet sufficiently good knowledge about atoms and molecules, yet Thompson gave a great deal of consideration to what he called "the constituent particles of all bodies" and their motion, and thought that "these particles must be so extremely small, compared to the spaces they occupy, that there must be ample room for all kinds of motion among them."

Thompson reported his study in papers titled "The Propagation of Heat in Various Substances" and "New Experiments Upon Heat." Some of his results posed no threat to the caloric theory, but others were serious threats. He proved that the caloric theory's prediction that heat could not be conducted through a vacuum was false, and that "the heat generated by friction is *inexhaustible*. . . . It appears to me that that which any insulate body or system of bodies can continue to give off without limitation, cannot be a *material substance.*" In "Sources of Heat which is Excited by Friction," he wrote specifically of his discoveries resulting from his observations of the boring of cannons. So that other experimenters without access to an arsenal could repeat his findings, he designed a device to demonstrate the production of heat through friction in a simpler way.

Forty years later, when the English scientist J. P. Joule formulated the first law of thermodynamics, he credited Thompson with making the first measurement and estimate of the mechanical equivalent of heat, addressing the question of what amount of mechanical work would raise the temperature of a quantity of water to a certain point: "According to Count Rumford's experiment," Joule wrote, referring to Thompson's experiment with the horses, the brass cylinder, and the water, "the heat required to raise one pound of water one degree will be equivalent to the force represented by 1034 foot-pounds."*

Though Thompson knew that these discoveries were highly significant, he struck a humbler tone in an essay identified only as "Thompson Essay VII" in biographer Sanborn C. Brown's *Benjamin Thompson, Count Rumford*. After composing a condescending sentence about the fear of God possibly helping people of "more barbarous nations" enjoy peace and prosperity, Thompson went on to write, "belief in the existence of a Supreme Intelligence who rules and governs the universe with wisdom and goodness, is not less essential to the happiness of those who, by cultivating their mental powers, have learned to know how little can be known." He told a friend that this sentence "I conceive to be of more importance than any I ever penned."

Thompson's discoveries and papers again brought accolades for him and prestige to the Bavarian court, but they inevitably aroused yet more enmity. When Theodor decided to resolve an unpleasant situation for both himself and Thompson by naming him Minister Plenipotentiary to the English court, the well-meant move led to a humiliating disaster. Thompson and Sally arrived in London in September 1798 and were greeted in the lobby of their hotel with a letter telling him that the Court of St. James could not under any circumstances receive him as Bavarian minister. The elector had neglected to obtain King George III's permission. The official explanation was that, as a British

---

* A foot-pound is a unit of work equal to the work accomplished by a force of one pound acting through a distance of one foot in the direction of the force. One foot-pound is equivalent to 1.35582 joules.

subject, Thompson could not represent a foreign government at court, but Thompson suspected that his enemies had succeeded in making a fool of him.

Discouraged and at loose ends, Thompson considered returning to America and sent Sally ahead, but when the Royal Society urged him to write more about his heat experiments, that was enough to keep him in London. He was soon engaged in the project that would be his second significant contribution to science: founding what would become known as the Royal Institution of Great Britain. Thompson planned it as a museum of science—specifically, practical mechanics—with model rooms demonstrating kitchen and ventilation improvements, a laboratory of apparatuses, and lecture rooms. The object was to provide technical instruction that would allow poor laborers to be more effective mechanics and artisans. As time passed and Thompson struggled to raise funds and placate others involved in the Institution's founding, that vision faded to resemble something more like a meeting place where the upper classes and intellectuals could learn about experiments and sciences. A frustrated Thompson was unconvinced by arguments that such knowledge would filter down to benefit the lower ranks of society. He grew increasingly dictatorial. Eighteenth-century political commentator John Wolcot, who wrote under the pen name "Peter Pindar," described him as a man who was contemptuous of others' opinions and who asserted that "I never was yet in the wrong; I know everything." Even if that were true, "Pindar" continued, such a declaration was "arrogant and supercilious." A month before the Royal Institution officially opened on March 4, 1800, Thompson became so obsessed with getting everything right and so distrustful of the abilities and motives of everyone else involved that he moved into the building to take full charge of even the minutest details of construction, remodeling, and purchases.

Professor Thomas Garnett, who had founded a similar institution in Glasgow, was invited—but not by Thompson—to be one of the first scientists associated with the Royal Institution. After the opening, he proceeded to deliver six lectures a week. These were a great success, especially among fashionable audiences, but he and Thompson were at odds from the start, largely because others had made the decision to offer Garnett the professorship while Thompson was away. To lecture on chemistry, Thompson hired Humphry Davy. This choice was probably made more to upstage Garnett and have a man more under his own thumb than for Davy's science, although Davy turned out later to have been a brilliant hire. The functions dearest to Thompson's heart—lectures for mechanics and laborers and kitchen exhibits—were slow coming to fruition. There was opposition from the industrial establishment to exhibits from which some of their most valuable innovations could be copied (the patents had run out). This particular clash ran deep, for Thompson felt strongly that no inventions or science should remain privately owned or withheld from copying and wider use. He never patented his own inventions.

This cartoon by James Gillray, entitled "Scientific Researches! New Discoveries in Pneumaticks! Or, an Experimental Lecture on the Powers of Air" (1802), satirizes a lecture on pneumatics given by Humphry Davy at the Royal Institution in London. Standing behind the table are Thomas Garnett (administering nitrous oxide, or laughing gas, to John Hippisley, one of the corporate managers of the Royal Institution) and Humphry Davy (holding the gas pump). Benjamin Thompson is shown standing on the top right in the blue coat.

Thompson's relationships at the Institution became increasingly troubled. His opposition to Garnett rose to the level of persecution and probably contributed to the decline that ended in Garnett's early death in 1802. In spite of the success of his lectures, Garnett's position was downgraded and his pay was never increased, while others received raises. Eventually Garnett resigned in discouragement. Thompson also was discouraged—the Royal Institution was moving increasingly far away from his beloved plan.

In September 1801, Thompson set out traveling again. In Munich, he was warmly received by the court, his former mistresses, and his daughter Sophy. He visited his "English Garden" and enjoyed himself more than he had expected in the city where he still had many enemies. He continued to Paris, where he was presented to Napoleon. The French military leader took an enormous liking to him. Anyone on whom Napoleon was showering such attention attracted similar attention from a great many others. Thompson was a sensation and in his element. He met mathematician Pierre-Simon de Laplace and mathematician-astronomer Joseph-Louis Lagrange—two men whose reputations have endured much longer than his—but he was too late to meet the chemist Antoine Lavoisier, who had been guillotined in 1794. It was Lavoisier's widow, Marie-Anne Paulze Lavoisier, who so captured Thompson's fancy that what was intended to be a two-week visit lasted two months. After a brief return to London to wrap up his affairs, in 1802 he left England permanently and moved to Paris.

The French academic establishment and French society treated Thompson with respect amounting to adulation, and he continued his affair with Marie-Anne Lavoisier. She had taken an active part in her husband's scientific work during his lifetime and was highly regarded in Paris social and academic circles. After Lavoisier's execution, she had received his possessions and a note saying that the charges against him had proved false. With his name cleared, she began to host a prestigious scientific salon and had several suitors, but it was Thompson she preferred. Lavoisier had given the "caloric" theory of heat its name, and the irony was not lost on Thompson. He had defeated the caloric theory and now had won Lavoisier's wife. Thompson made inquiries and learned that his wife in America had died in 1792, which left him free to wed Marie-Anne. Bizarrely, what had been an extraordinarily happy four-year love affair dissolved into ugly disputes when the couple married in 1805. Their marital difficulties and the petty malicious mischief they made for one another became the talk of Paris. The marriage ended in a formal separation after only a few months. Afterwards, the couple found it possible to relate to one another again in a civilized manner. They were legally divorced in 1809.

Rumford took a house at Auteuil, a suburb of Paris, and there at last devoted himself entirely to his scientific interests. He invented a drip coffee pot and continued to write papers and compose his memoirs. His daughter Sally returned from America to join him in 1811. It was at Auteuil that Thompson took his last mistress, Victoire Lefèvre. She bore him a son, Charles François Robert Lefèvre, in 1813. Sally absented herself during the pregnancy and birth but returned to care for her father and his household until his death.

Not long before he died, Thompson had a visit from Humphry Davy and the young Michael Faraday, Davy's assistant at the Royal Institution. The internationally acclaimed work of these two scientists would be instrumental in giving Thompson's Royal Institution the prestige that it still enjoys today. When it seemed likely that the Russian army would lay siege

Antoine-Laurent Lavoisier and His Wife, *1788, a portrait by French painter Jacques Louis David depicting chemists and spouses Lavoisier and Marie-Anne-Pierrette Paulze. Marie and Thompson married in 1805, only to separate after several months.*

to Paris after the fall of Napoleon, Thompson sent his daughter away for safety, and she had not yet returned when he died suddenly in August 1814, at age sixty-one. His divorced wife was also away. Thompson's funeral was an obscure, sparsely attended event in the cemetery of Auteuil.

Thompson had not patented any of his practical inventions, including his fireplace and kitchen devices, on the argument that such things ought to be free for anyone's use. Nevertheless, he left his daughter well provided for. He donated all his books to the United States government to form part of a military academy library whenever such an academy would be founded. The residue of his estate went to Harvard College.

In spite of Joule's recognition of his contributions to the understanding of the nature of heat and his numerous practical inventions, Thompson was to become hardly more than a footnote in the history of science. The work that disproved the caloric theory was, indeed, too far ahead of its time. However, in the introduction to a compilation of essays titled *The Correlation and Conservation of Forces: A Series of Expositions*, published in 1865, American scientific writer Edward L. Youmans raised the possibility that Thompson's obscurity in the history of science was to some extent contrived by those who disliked him as a person or sought to promote strictly "English science."

> *There has been a manifest disposition in various quarters to obscure and depreciate [Thompson's discoveries]. Dr. Whewell, in his* History of the Inductive Sciences, *treats the subject of thermotics without mentioning him. An eminent Edinburgh professor, writing recently in the* Philosophical Magazine, *under the confessed influence of "patriotism" undertakes to make the dynamical theory of heat an English monopoly.*

Thompson is remembered today in America not because he made a significant scientific discovery but as a donor to Harvard College. There is still at Harvard a "Rumford Professorship of the Physical and Mathematical sciences as applied to the useful Arts," but those appointed to the professorship may very well need to look Rumford up to find out who their benefactor actually was.

THREE

JEAN-BAPTISTE CHAPPE D'AUTEROCHE

PURSUING VENUS

(1677–1770)

N NOVEMBER 7, 1677, THE DAY BEFORE his twenty-first birthday, Edmond Halley watched a black dot cross the face of the sun. It was a "transit" of the planet Mercury. Halley was on the island of St. Helena in the South Atlantic, engaged in mapping the stars visible from the Southern Hemisphere, but he took the time to ponder how such a transit could provide information that would make it possible to measure the distance to the sun. Halley knew that fourteen years earlier, Scottish astronomer Rev. James Gregory had suggested viewing a transit of Mercury or Venus (the only two planets that ever pass between Earth and the sun) from widely separated locations on Earth, to find their "parallax." That measurement, and the proportions of the solar system known since Johannes Kepler discovered his third law of planetary motion in 1618, would allow scholars to calculate the Earth-to-sun distance.

Transits of Venus are rarer than transits of Mercury. There had been one in 1639, before Halley's birth, and he calculated that there would be another

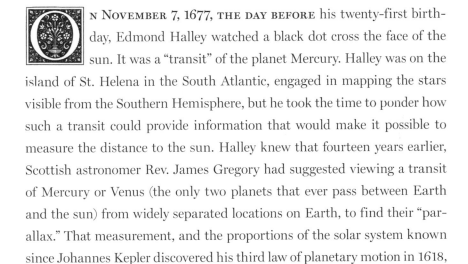

on June 6, 1761. Not hopeful of living to age 105, he published detailed instructions for using data from widely separated sites, where observers would see the planet touch the sun's disk at different times. Halley's invitation to future astronomers ended with a prediction that this work would "redound to their immortal glory."

OPPOSITE: *A c. 1888 engraving from a French weekly,* La Science illustrée, *depicting Jean-Baptiste Chappe d'Auteroche during a lightning storm in Tobolsk, Siberia, while observing the transit of Venus there in 1761.* ABOVE LEFT: *An engraving of English astronomer Edmond Halley, 1722.*

61

SIMPLE PARALLAX DEMONSTRATION

For the simplest demonstration of a parallax shift, hold one finger a short distance before your eyes, focus on the background, and close first one eye, then the other. Your finger seems to shift from side to side against the background. The farther away you place your finger, the smaller the shift. In Gregory's and Halley's plan, two distant places on Earth would act as the two "eyes." The black dot of the planet would be the finger, the sun the background. We don't usually stop to calculate precisely how far our finger is from our eyes—our brains do that satisfactorily for all normal purposes—but it is possible, from this shift, to calculate the distance to Mercury or Venus, the two planets that pass between Earth and sun.

Eighty years later, as the year 1761 approached, teams of observers prepared to take up Halley's challenge. Kings diverted money from wars. Locations were calculated and debated. Months-long, dangerous voyages and overland treks began. There are colorful stories connected with this international undertaking—life-risking escapades that sound more fit for Indiana Jones than the pursuits of scientists—all in hope of observations that would fail if a cloud passed over the sun at the wrong time. French clergyman and astronomer Jean-Baptiste Chappe d'Auteroche was the most intrepid of them all, and most successful. He was also one of the unfortunate, courageous, foolhardy few who have died for science.

Born to the nobility in the Auvergne region of France, Chappe didn't need formal education or a profession to support himself. From childhood he excelled in mathematical calculations and had a passion for studying the stars. He worked as a surveyor for aristocratic clients when it suited him, but mostly he watched the night sky. He must have made considerable progress

Chappe, depicted in an engraving from the Literary Magazine and British Review, *1791.*

developing his skills as an astronomer, for in 1759, at age thirty-seven, he became an assistant astronomer at the French Royal Observatory and was elected to the French Académie Royale des Sciences. A year later he would begin his travel to observe the 1761 Venus transit. A colleague described him as a "candid, unpretentious, straightforward" man, "naturally lively, gregarious and amiable." In the early 1760s, Chappe was no eminent elderly astronomer but a younger, stronger, and perhaps more expendable man.

He wasn't the Académie's first choice for a Venus assignment—the more distinguished Guillaume Joseph Hyacinthe Jean-Baptiste Le Gentil would be observing from Pondicherry, India—but, as it happened, an unexpected overture came from the Russian Imperial Academy of Sciences and Arts in Saint Petersburg. Not to be outdone by the rest of the scholarly world, they had urged Russian astronomers to step forward for an expedition to Siberia. None had volunteered, and it had become necessary to sponsor a foreigner. When Chappe embarked on the four-thousand-mile journey from Paris to Tobolsk, Siberia, he did so believing he was sponsored by both the French and Russian academies.

His destination was no afterthought. Chappe's colleague at the Académie, astronomer Joseph-Nicolas Delisle—age seventy-two and no longer able to make long journeys himself—had presented the Académie with a map of the world, a "mappemonde," pinpointing the best destinations for teams intending to study the Venus transit. The most valuable information would come from observations at the transit's "extremes," that is, where its duration was longest and shortest. Tobolsk, where the transit would be shortest, was a choice observational location but extremely remote, devilishly hard to reach, and poorly mapped. Adding to the difficulty, Russia discouraged visitors.

The invitation from Saint Petersburg was a rare exception and one that King Louis XV of France recognized as a not-to-be-missed opportunity to learn more about Russia. Chappe was instructed to keep careful notes. Later, he began his book *A Journey into Siberia* with the words "Being ordered by the King and appointed by the Académie. . . ." The king was listed first.

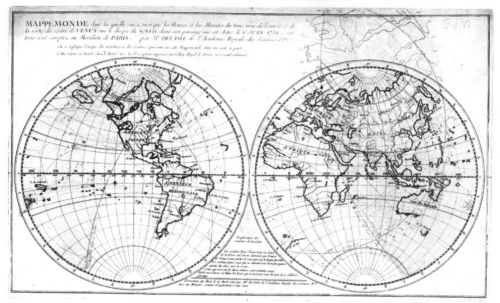

Mappemonde *by French astronomer and cartographer Joseph-Nicolas Delisle tracing the path on Earth from which the 1761 Venus transit was observable. The faint, curved line can be seen running through Siberia.*

Chappe intended to begin his journey in November 1760 by sailing from Holland to Saint Petersburg, but delays kept him in Paris beyond the ship's last possible departure date that autumn. He would have to travel overland instead, by way of Vienna and Warsaw, carrying with him his "large apparatus of instruments." The route was longer and more difficult, and the transit date was fast approaching, but regrets about missing the boat evaporated when he learned it had been wrecked off the coast of Sweden.

Chappe's lengthy, vividly written journal is a veritable catalogue of the discomforts and hazards facing travelers even within the relatively civilized confines of late-1700s Europe. The roads to Strasbourg were so sodden that what would in the best of weather have been a two- or three-day trip took eight days, leaving his carriages damaged beyond repair and his barometers and thermometers broken. After replacing them he altered his plans again, expecting roads in Germany to be even worse. The party would travel to the city of Ulm and take the Danube River to Vienna, although "navigation of this river was very uncertain at [that] time of year on account of the fogs" and the water was running so high that boats could fit under bridges only if great care was taken to avoid the most rapid current. Passages between projecting rocks were so narrow that vessels traveling in opposite directions couldn't pass one another and so lengthy that "there [could] be no hopes of preservation, if a shipwreck should happen." Travel was slow, navigation at night impossible. Chappe used the enforced stops to venture on foot into the mountains, measure their heights, glimpse the taller mountains beyond, and notice the extent of snow cover.

Some adventures early on had nothing to do with the conditions of roads or rivers. Chappe discovered a stone with a mysterious inscription and was disappointed to learn from local peasants that it was not a primitive relic but a recent Jewish tombstone with an epitaph in Hebrew. He and his companions saved a young man who, after a row with his mistress, was about to throw himself into the Danube. They rescued a lady whose uncle was forcing her to take the veil. They lost all of Chappe's underwear in a robbery while attending a Christmas Eve service.

The expedition arrived in Vienna on December 31 and celebrated the New Year with the glittering society and royalty of the Habsburg Empire, then basking in one of its happiest moments under the rule of Empress Maria Theresa and her husband, Franz I. Chappe met Jesuit Father Maximilian Hell, who would observe the transit from Vienna—a much more comfortable site than where Chappe was headed.

After Vienna, Chappe and his company left the Danube and began to encounter difficulties that would plague them repeatedly for the entire journey: carriages, sledges and equipment broken in ditches or by the rough surface of the lanes and cold such as they had never before experienced.

Rivers became major obstacles. As winter set in, ferries stopped operation, but the ice was still too thin to support horses and carriages. To ford shallow streams, ice near the banks had to be broken to get in and out, and loose floating ice made footing difficult for the horses. Where they entered Poland, reaching the Vistula River, Chappe ordered the thin ice broken so that the ferry could take them across.

January 22: They reached Warsaw. Chappe's detailed notes called it a "very fine city, [with] several elegant buildings . . . [but] not one single inn." Fortunately, "the Polanders are indeed so very hospitable" that no one was left out on the streets. The women were "in general handsome and amicable," could read and knew various languages, and dressed in the French manner except for casual dress, which was more Polish, "sort of very elegant riding-dress." They "love company and pleasure, but are strictly virtuous," the effect either of "superstition . . . the climate, or of true principles of religion." It was the season of masked balls preceding Lent.

Chappe enjoyed Warsaw but found Poland a nation, due to the complete independence of the nobles, "subjected to the neighboring powers . . . [leaving] the state without defense, and exposed to every invader." There were land-mongering aristocrats, laws of inheritance that could

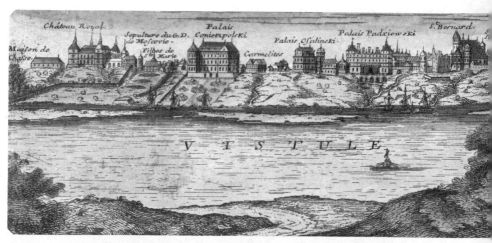

A panoramic map of Warsaw featuring the Vistula River in the foreground by French cartographer Nicolas de Fer, 1705. A half-century later, Chappe and his expedition would cross the river by ferry during their travels to follow Venus.

render a wealthy man a pauper in a day but still allow him considerable power, virtual slavery of the country people, and harsh punishments for trivial offenses. Of most immediate concern were a robbery and multiple murders that had recently occurred on the road Chappe was about to take.

January 27: The expedition departed Warsaw with the Vistula frozen enough to cross. After a pause at the vast, elegant estate of M. Branisky, Grand Marshal of Poland, they found the roads narrower and the cold more intense, and there was still too little snow to use sledges. This was one of the most difficult segments of the journey, though the expedition hadn't yet even left Poland. The hills, though not steep, were covered in ice. Chappe's ten Polish horses, not shod, all had to be attached to a carriage for an ascent, then, at the summit, harnessed to the back of it to brake the descent. When the ice in a stream gave way under one carriage. Chappe borrowed four local horses late at night and harnessed them along with his ten to pull it out.

At Riga, in Latvia—still well short of Saint Petersburg—Chappe acquired sledges that were convertible to wheeled carriages. This was a wise move since the sledges began to scrape over snowless ground soon

after leaving the town. Chappe and his companions struggled to make the conversion from sledge to carriage, which was no easy task with the loads so heavy and the cold so severe. Near Wolmar they were traveling again on snowy roads when the lead carriage vanished into a sinkhole deep enough to conceal it completely except for the horses' frantic heads poking up above the snow. "We were buried in the carriage; there remained only a small opening at the top of the coach by which we got out." After two hours spent extricating themselves and the horses and carriage, they converted the carriages back to sledges. That afternoon at Derpt, Russian travelers coming in the other direction warned them they were to face still-deeper snow and narrower roads. Chappe dispensed with wheels altogether and bought four sledges.

February 13: Just short of three months after leaving Paris and still 1,800 miles from his destination in Siberia, Chappe arrived in Saint Petersburg and learned that the Russian Academy had grown less enthusiastic about

A colored engraving entitled A View of St. Petersburg on either side of the River Neva *(1750); the Russian Academy of Sciences is shown on the left. Chappe visited the Academy in February 1761, but its members were dubious about Chappe's ability to reach Tolbolsk in time to witness the transit.*

Portrait of Empress Elizabeth of Russia, a Chappe patron, by Austro-Italian painter Georg Caspar von Prenner, c. 1754.

supporting him. They suggested that he watch the transit from nearer Saint Petersburg. Chappe argued that the transit as observed from Tobolsk would be the shortest of anywhere in the world—a detail of vital importance. French ambassador Baron de Breteuil supported him, as did the high chancellor of Russia, Count Woronzof, whom Chappe described as "a lover and protector of the Sciences." Chappe gave even more effusive thanks to the Empress Elizabeth, who "gave the most circumstantial orders on this occasion. All the assistances I met with in travelling through Siberia were owing to the protection of this Princess." One could not do better for patronage.

March 10: After a fraught four-week delay while all this was sorted out, the expedition set off into an area where provisions, "even those which are most in common use," were unavailable. If the company wanted bread or beds, they would have to carry them. An interpreter/guide and a servant joined the group, along with a clock-maker, for Chappe's instruments were continuing to suffer mishaps. The party traveled in four sledges, each drawn by horses harnessed five abreast. They paused only for changes of horses. After each break, Chappe usually insisted on getting under way again before dusk so that travel continued day and night. He'd become accustomed to bone-jarring sleep in his sledge, waked each time he was abruptly thrown off into the snow. The sledges were unwieldy and prone to overturning, but Chappe grew fond of them. Between disasters, they whisked him over the frozen ground and iced-over waterways at a faster pace than most travel took place in his century. The speed was welcome. Chappe's greatest fear was that the looming thaw would render sledges useless and make river ice so thin that it would be impossible to cross.

March 14: Chappe reached Moscow with his sledges damaged beyond repair. A letter he had brought from Count Woronzof to Woronzof's brother in Moscow had replacements immediately under way, but when it was predicted that the spring thaw would almost certainly be complete before the expedition could possibly reach Tobolsk, Chappe decided not to wait. He bought used sledges and set out immediately.

At first, travel after Moscow went swiftly. Chappe even stopped to investigate and sketch curious layers of various colors in the mountainsides, "resembling at a distance a brick wall." He waded into deep snow, hazarding entrapment in a crevice, and sketched the "wall." On the rivers, his curiosity was piqued by large unfrozen holes in ice that was otherwise three feet (.91 meters) thick. A horse plunged into one, and Chappe saved the others harnessed to that sledge by quickly cutting the attaching cords.

The Volga River was particularly beautiful where it joined the river Occa at Nizan-Novogorod (modern-day Nizhny Novgorod), and there were fish and meats in the market that had been frozen for four or five months. His report about the people, however, was less positive. Boys married at fourteen or fifteen and girls at thirteen, "in order to prevent debauchery." Women gave birth until middle age and beyond. Chappe met a forty-year-old postmaster's wife who'd had twenty children, only two of whom had survived early childhood.

After Nizan-Novogorod, the frozen Volga became their road, the sledges traveling "with inconceivable swiftness." Chappe, remembering the holes in the ice, suggested they not travel at night, but his guides assured him they knew where the holes were. He "sat as easy in my sledge on the Volga as I should have done in a boat in the summer time." He stood upright for a while, enjoying other traffic on the river, but "the severity of the cold obliged me to get quickly down into my sledge again, neither could I bear the excessive quickness of the motion while I stood upright."

The horses, though small, were fast. The drivers seldom used a lash but communicated with them by whistling, waving their hands, and talking to them in "baby talk."

At Kozmodemyansk, they left the river behind and entered the forest, not to emerge from it again until they were almost at Tobolsk. The roads were so narrow that if two sledges met, one was laid on its side to let the other pass. Chappe's first horse had a bell attached to signal approaching parties that theirs was a "royal post" and must be given the road. Royal post or no, it was difficult to find fresh horses and local helpers. Chappe had been hiring "postilions," men familiar with the road who would ride one of the lead horses, but here the local men fled into the forest as Chappe's party approached. Women and babies cowered and hid. Chappe learned that "most travelers made free with the horses, and with every thing else belonging to the inhabitants of these hamlets." Villagers were "ill-used upon asking for what was their due" for food, horses, and riding postilion.

Chappe described miserable living conditions in windowless hovels heated by filling them with smoke—the results of which were either unbearably hot or unbearably cold depending on the time of day and the state of the fire. For nine months of the year the people seldom went outside. The smell was ghastly. The villagers were devout Greek Orthodox, and each home had a "chapel" with images and perpetually burning small candles that sometimes caused fires—infernos that consumed houses and even entire villages.

One night, Chappe awoke, near frozen, to find himself alone with no light or sign of the rest of his party or the other sledges. He attempted to search the deep, snowy forest in the darkness, stumbling here and there, getting confused about the way back. He slipped and tumbled into the snow, eventually struggling out only to find that he had lost the small path he had been following. Apparently abandoned with nothing around him but hundreds of miles of forest, Chappe summoned the courage to set out on foot one last time. It wasn't long before he spotted a dim light in the distance—his party hadn't deserted him. They were asleep in a cottage in the arms of young girls, having "stopped longer than they intended."

In his journal, Chappe recorded instructions how to bait and hunt black bears and polar bears but didn't attempt this himself. At Solikamsky he tried a sauna, "unwilling to leave the country without being convinced by my own experience of what had been reported to me concerning these baths." He found the heat suffocating, the flogging with twigs miserable, rubbing his body with onions and rolling in the snow after his bath totally unappealing. Only later, after learning better how it should be done, did he find the baths not intolerable but never enjoyable.

At Verkhoturye, Chappe's expedition finally entered Siberia and turned south, where the forest gave way to more open country and larger villages. It became increasingly difficult to persuade his drivers to risk crossing frozen rivers now ominously coated with a sheen of water. At Tyumen they balked completely. The ice was expected to break up any minute, and Chappe was urged to wait until morning. He refused to risk the delay, fearing that by morning boat travel would be impossible as the thaw swelled the river and poured torrents down its banks. Chappe offered his men double wages and gave them so much brandy to drink that they finally agreed to cross, with the ice groaning and water lapping the horses' hooves.

Seventy-five miles short of Tobolsk, the risk was even greater. Not one man would hazard his own horses. In answer, Chappe bought the horses and set up a thermometer so that his men could watch the temperature drop below the mark where the river could be deemed safe to cross. Chappe wrote that one man reported "with enthusiasm, that 'the animal' had got down below the mark," and he stopped his interpreter from explaining that the mercury was not an animal. This was not the moment to pause for a science lesson. Approaching the river in the darkness, they saw that so much water was flowing over the ice that the stars were reflected in it.

The first postilion was preparing to cross it, and stopped short. I stood upright on my sledge; and called out to him [to go on]; pushing, at the same time, my own postilion so violently, that he went on immediately. The first postilion, not willing to be overtaken, gets on at a still greater rate; the others follow, and we were on the other side of the river in an instant.

Loading sledges and horses with supplies during the Siberian expedition; from Voyage en Sibérie.

On the opposite bank, Chappe was mysteriously seized with convulsions, tremors, and complete loss of strength. He drank some liqueur, felt better, and fell asleep for a good part of the remaining journey. The river Irtysh still lay between him and Tobolsk and could have been the final deadly barrier, but people were happily crossing its still-frozen surface when Chappe arrived on April 10, six days before the ice finally broke up. He had traveled the 1,800 miles from Saint Petersburg in one month.

Chappe recovered his strength rapidly and hastened to prepare for the June 6 transit, setting up the instruments that had survived—"a quadrant, clocks, a parallactic machine, and a telescope nineteen feet [5.8 meters] long." For his observatory site, he chose a mountain three-quarters of a mile from Tobolsk with a view of the entire horizon.

The days grew longer; the nights shorter. Chappe took preparatory observations of the sun, moon, and stars, and fine-tuned his quadrant. No one had told the local people what Chappe was about, so they grew suspicious and fearful seeing him return to town each morning at about 4:30 a.m. Surely this haggard, shabbily dressed man was a magician? Some heard whispers of "June 6" and concluded that the end of the world was expected that day. To make matters worse, the Irtysh River rose disastrously high. Townspeople watched houses in the lower town disappear in water up to the roofs and neighbors being swept away in their collapsing homes or in the streets while trying to rescue their belongings. Even part of Chappe's observation mountain plunged into the river—higher elevations on the plain became scattered islands on a body of water that stretched farther than the eye could see. The frightened populace were was convinced that the floods would not recede until the world ended . . . or until Chappe left.

Tobolsk's governor thought it prudent to supply Chappe with guards as the uneasiness grew, but Chappe, absorbed in his work, didn't at first realize that the local antagonism made it essential that he take these men with him as he went to and from his observatory. Only when the hostility became impossible to ignore did he make sure they accompanied him. He spent the nights in the observatory fearing a mob would pull it down.

May 11: Chappe had his apparatus ready to observe an eclipse of the moon that would occur on May 18, an observation necessary to establish the exact longitude of Tobolsk.

June 4: Two days before the transit, a strong wind almost carried away Chappe's observatory. It died down by noon the next day.

June 5: With one day to go, Chappe set up a tent with a telescope to keep the local dignitaries and their families—who planned to gather nearby and watch the phenomenon—distracted so that he wouldn't be disturbed during his observations. By evening, all was in order. Then, at 10:00 p.m. fog and clouds began to gather and thicken. "The bright sky disappeared; and the whole hemisphere was soon overspread with one single black cloud." With the scientific world holding its breath in anticipation of the next morning, it was almost impossible to bear

> *the idea of returning to France, after a fruitless voyage; of having exposed myself in vain to a variety of dangers, and to fatigues, under which I was supported only by the earnestness and expectation of success, which I was now deprived of by a cloud. . . .*

June 6: The sun rose, and an easterly wind drove the clouds toward the west. They turned whitish and brighter and began to disperse, and with them went Chappe's gloom. The weather was so promising—"the calmness and serenity of the air"—that Chappe brought his instruments out of the observatory and into the open. The dignitaries arrived; the guard proved unnecessary because most townspeople were hiding in the churches and houses awaiting the end of the world. Chappe's watch-maker was poised to keep an eye on the clock and take notes. The interpreter would count the time. The moment was finally at hand.

The clouds dispersed sufficiently for Chappe to see one border of the sun, but Venus was to enter on the opposite side. By the time those clouds had disappeared, Chappe saw that Venus had already touched the sun's disk. However, not all opportunity to gather data was missed—the second important observation would be of the exact moment when Venus had totally entered the disk of the sun. Chappe was "seized with an [*sic*] universal shivering, and was obliged to collect all my thoughts, in order not to miss it." He was able to make that observation successfully, recording it at twenty-eight seconds past 7:00 a.m. The sky remained clear, and Chappe recorded both the last moment when Venus was entirely immersed in the sun's disk and the moment when it last touched the sun's rim. He submitted his report to Saint Petersburg a few days later by a dispatch that the governor of Tobolsk sent regularly to the Russian court.

Chappe's report was published in Saint Petersburg almost as soon as it arrived there. He had expected it to be forwarded immediately to the French Académie, but it wouldn't reach Paris until a year later, after he had returned there himself in August 1762.

Meanwhile, Chappe's return journey from Tobolsk, begun August 28, 1761, took place in a warmer season and was not so harrowed and frantically hurried as his trip out. He reached Saint Petersburg on November 1. In January 1762, when he delivered a lecture about his transit observation at the Russian Imperial Academy of Sciences and Arts, the buzz in the distinguished, glamorous audience as they awaited his appearance was possibly more focused on the recent death of the Empress Elizabeth and the chilling concern over having her nephew, Peter III, as ruler. Chappe spoke eloquently about the success of his expedition and the glory it reflected on the Russian court that had sponsored it, but he declined that an offer had come from the Empress to make him Imperial Astronomer.

Even before Chappe left Tobolsk in the late summer of 1761, reports of observations from colleagues whose posts had been nearer to home were circulating all over Europe. Scientists and academies began to collate and study the measurements, and it was becoming clear to those analyzing the results from no fewer than one hundred twenty teams that the project hadn't been as successful as hoped. The three most important pieces of information that each team reported were the length of time it took Venus to cross the sun, the precise latitude and longitude from which they had observed, and the method used to arrive at those numbers. Unfortunately, the exact geographical positions of some of the observation posts hadn't been established with the same meticulous care that Chappe exercised. Even more troublesome was the fact that most of the observers reported that, as Venus approached and just touched the border of the sun's disk, the planet had seemed to stall for a moment, its image distorted. From Madras (Chennai) in India, it had looked like a pear; from Uppsala, Sweden, like a drop of water or the tip of a rapier.

The distortion would become known as the "black drop effect," an optical phenomenon that made it impossible to time precisely the moment when the planet first touched the disk of the sun and when it left—more technically, the *ingress* and *egress*. Chappe hadn't observed the ingress of Venus because of the cloud cover of that edge of the sun's disk, and he never reported anything odd about the egress. It is possible that something about Tobolsk's atmosphere saved his observations from the anomaly others had experienced. On the positive side, earlier Earth-to-sun measurements hadn't been far off. The French expeditions' reports, analyzed together, gave an Earth-to-sun distance of 85.3 million miles. The best measurements previous to 1761, made nearly a hundred years earlier by Jean-Dominique Cassini (Gian Domenico Cassini) in 1672, put the distance at 87 million miles.

Nevertheless, all this taken into consideration, it was a very good thing that another Venus transit was predicted to take place on June 3, 1769.

The path of a Venus transit. On June 5–6, 2012, NASA's Solar Dynamics Observatory, or SDO, collected images of a transit of Venus across the face of the sun. This rare event happens in pairs eight years apart, and those pairs are separated from each other by 105 or 121 years. The next transit will not happen until 2117.

Chappe didn't publicly share his reasons for declining the Russian Empress's offer to make him Imperial Astronomer, but one explanation is surely that he was anticipating observing this second transit, and no Russian location would be ideal for that. Another explanation is that, in spite of all his fine words about Russian sponsorship, Chappe was not fond of Russia—an attitude that wouldn't be revealed until his journal was published.

In 1768, on the eve of Chappe's second Venus transit expedition, the account of his Siberian journey written for Louis XV finally made its public appearance. The book was a splendid publication, illustrated with numerous maps and fine drawings and prints by Jean-Baptiste Le Prince from his own travels in Russia, but it was not a pleasant portrait of Russia. Chappe wrote about the despotic manner in which Russia and Siberia were ruled, their history, their politics and taxation, the discouragement of education,

Illustration from Voyage en Sibérie *depicting a peasant being whipped with a knout—a brutal type of rawhide whip sometimes ending in wire hooks and often used in Imperial Russia for corporal punishment.*

arts, and sciences, and the lack, even among the military, of expertise in the art of war. He described "cruel and inhuman" punishments for the most trivial offenses, which he'd witnessed firsthand. He reported on the naval and land forces of Russia—with long lists and names of her vessels and analyses of the state of her troops, their pay schedule, and the rate of desertion—and her import and export trade with foreign countries. He presented a devastating picture of the "common people," the men given to drunkenness and debauchery, squandering whatever money they had on drink, with laziness "the greatest pleasure they have." Young peasant women had only one means of amusement: riding on a seesaw—something new to Chappe.

Six years earlier, in July 1762, a conspiracy in Russia led by Peter III's wife—soon to become known as Catherine the Great—had overthrown him and crowned her. Chappe's beautiful book enraged Catherine. His careful records and observations were a shocking contrast to the favorable image of Russia she was trying to project abroad. Chappe had already departed to pursue the next transit of Venus when a scathing pamphlet appeared. Two hundred vitriolic, comical pages made fun of Chappe, insinuating that he had an unhealthy interest in peasant women, ridiculing his claims to read a thermometer he had previously said was broken, and pondering how he could make so many geological observations and observe so many women while cowering from the cold in an enclosed sledge traveling at top speed. The author was anonymous, but rumor had it that it was written by Catherine herself. Chappe probably never saw the pamphlet.

At a meeting of the Académie Royale des Sciences in Paris in November 1767, Chappe proposed an expedition to the South Pacific to observe the next transit of Venus, in 1769. The Spanish were laying claim to islands there and refused permission, but Carlos III of Spain was willing to offer passage on a vessel sailing with the Spanish fleet to Mexico. One of the best locations for viewing the transit would be the peninsula on the Pacific Ocean, now known as Baja California. The French Académie assigned Chappe to make the journey across the Atlantic Ocean, Mexico, and the Vermeille Sea (now the Gulf of California or the Sea of Cortés) to the tip of Baja California.

Voyaging to the Spanish city of Cádiz to join the fleet, Chappe took with him an assistant, Jean Pauly, identified as "the King's Engineer and Geographer"; Alexandre-Jean Noël, a sixteen-year-old painter and drafts-man, to record "sea coasts, plants, animals, and whatever we might meet with that was curious"; and M. Dubois, a watchmaker, to repair "the lit-tle mischiefs" the instruments would suffer along the way. Their baggage included five telescopes, two quadrants, the most up-to-date marine chro-nometer, a state-of-the-art compass, barometer, thermometer, and a new instrument to measure water density. Chappe had no instructions to report on a foreign country, but he intended to take careful notes and "draw other advantages from this voyage: that in case we should be so unfortunate as to fail in our main purpose we might in some measure make amends to the learned world for this loss."

After two months of bureaucratic stalling, the Spanish court gave Chappe permission to sail, not with the fleet but in a solitary boat with a crew of twelve. It could be suspected that Spanish authorities hoped to rid themselves of a troublesome Frenchman. Two Spanish naval officers, Don Vicente de Doz and Don Salvador de Medina, would accompany Chappe's party. Both were astronomers and would observe the transit from the same location as Chappe. The vessel, "so very light as to be the sport of the small-est wave," left Cádiz on December 21, 1768.

Chappe exclaimed in his journal about the courage of such men as Columbus, who had sailed "without a thousand helps that were wanting in the days." One wonders, reading his description of the ship and the voyage, whether sea travel had improved all that much. However, the "little nut-shell" was fast, and though Chappe found it impossible to make meaning-ful observations from the shifting deck, the voyage wasn't boring.

> I must say that the sea-faring life is tiresome and uniform to such
> only as have not accustomed themselves to look about them . . .
> but to an attentive spectator, the sea offers objects very capable of
> entertaining the mind, and exercising all the intellectual powers.

Nature has beauties even in her horrors; nay, it is there perhaps
that she is most admirable and sublime.

That was a healthy attitude for someone braving this voyage and about
to face the splendor and dangers of crossing Mexico.

After seventy-seven days at sea, they arrived outside the harbor of Veracruz
on the afternoon of the March 6. The ship's captain unwisely ignored the advice
of the two Spanish astronomers to fly the Spanish colors, and hence the ship
was greeted by gunshots when attempting to enter the harbor. Backing off, they
were stuck in a deadly position among rocks with a strong north wind blow-
ing—Veracruz was known as a dangerous harbor for good reason. Fortunately,
they were soon able to establish their identity and enter safer water, but by
that time the wind had become a hurricane and their vessel, with cargo still
onboard, was forced to screen itself behind the castle promontory, secured only
by cables. After three days of Chappe fretting over his instruments, the tempest
finally ended and the cargo made it safely to shore.

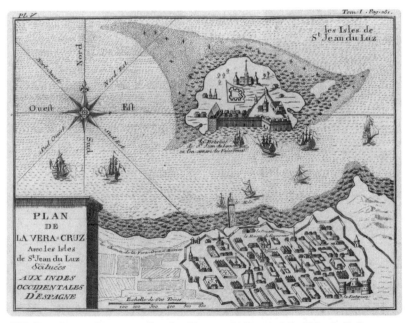

This French map from 1775, showing Veracruz harbor and the "Isles de St. Jean
de Luz" (the Spanish colonial fortress of San Juan de Ulúa), was created six years
after Chappe arrived there.

March 18: The eight-hundred-mile overland journey began on the same route along which Hernán Cortés and his army had traveled 250 years earlier en route to conquer the Aztec emperor Montezuma. Doz and Medina travelled in one litter, Pauly and Chappe in a second. Other members of the party rode mules, with a great many more mules carrying the baggage and driven by "Indians." Beds, tents, and provisions were part of the luggage, for there would be no lodging or food available for a good part of the trip, and no bread other than tortillas—a completely new fare to Chappe, who compared them to hardtack (sea biscuit). "As to the other messes that the Indians feast upon, they put in so much pimento, and pour such bad oil over them, that it is impossible, especially for a French-man, [*sic*] to touch a bit." After traversing sands and drained bogs that Chappe found "disagreeable and unwholesome," they arrived in a desiccated region where it was only possible to quench their thirst by paying peasant women they met on the road to take them to dairy cows.

In his journal *A Voyage to California*, Chappe described the dress of the Mexican people, their homes, and the near-slavery in which their Spanish conquerors held them. As in Russia, the women were little more than factories for breeding workers, this time for gold and silver mines. Girls were married at nine or ten years old and bore children until they were middle-aged. Few survived the two plagues—small pox and measles—that their conquerors had brought and that still ravaged a population with no immunity. Many perished in the mines. A race, said Chappe, was being destroyed, and the country as well, as lakes were drained and the province of Mexico became in places a desert compared to what it had been before the Spanish conquest.

The temperatures grew cooler as the expedition climbed above the clouds toward Mexico City. Even from the most elevated stretches of the road, the volcano Orizaba towered over them; "its top was wholly covered with snow, whilst the foot displayed the lovely verdure of rich cultivated land." At a village called Hapa they witnessed a Good Friday procession more like a "carnival masquerade than a religious ceremony."

March 26, Easter day: The expedition reached Mexico City and comfortable lodging in what had until recently been an establishment of the Catholic Jesuit order. They took meals every evening at the residence of the Marquis de Croix, Viceroy of Mexico. Chappe, though exceedingly critical of the ruling classes in Mexico and Siberia, never hesitated to take full advantage of their hospitality. He learned that the Jesuits had been expelled from this Spanish territory on pain of death for the Viceroy if even one remained, ostensibly for amassing too much wealth and threatening the supremacy of the Spanish crown. In fact, Chappe reported, they had become too sympathetic to the needs of the Indian peasants and had made too many successful efforts to establish schools to educate them.

Chappe's journal described Mexico City's churches and chapels, richly ornamented with solid gold and silver. The foot of the heavy Cathedral had already sunk six feet (1.83 meters) below the level of the plaza, for the Spanish city, Chappe learned, had been built on landfill. All of it had once been a magnificent lake surrounding the island on which Montezuma's palaces and temples stood. Chappe saw the *quemadero*, the execution site where "Jews and other unhappy victims" were still burned, "condemned, by judges professing a religion whose first precept is Charity." The Inquisition was so powerful that two years previously the viceroy, attending a meeting with them that he knew was going to end in orders to expel the Jesuits, took armed men along and surrounded the meeting place, giving instructions to destroy the place if he didn't emerge in fifteen minutes.

March 30: Chappe's party headed northwest out of Mexico City. He engaged an interpreter who spoke Spanish and also the Mexican languages and was acquainted with the country. The viceroy insisted he also take three soldiers, for the indigenous people in the territories they were about to cross were more savage, and rumored to have fabulous hoards of stolen gold, silver, and other riches secreted away in the forests. Chappe was warned that if they saw a man masked with a handkerchief, the safest response would be to kill him immediately. Expecting execrable roads, Chappe chose to proceed on horseback. Doz and Medina unwisely hired a carriage that suffered "a thousand mischances . . . which retarded us more than once." The baggage and instruments required forty mules.

The journey from Mexico City to San Blas, located on the west coast of Mexico somewhat south of Mazatlán and well north of Acapulco, was about 570 miles. The farther west the party traveled, the fewer people they met. The road was untended and dangerous, often built on the edge of precipices. The villages were desperately poor—Santiago de Querétaro was an exception, with its "noble," "very solid" aqueduct.

Near Querétaro, Chappe recorded a rare phenomenon: lightning rising from the ground to the sky instead of issuing from a cloud. On two occasions "we saw the same thing . . . and plainly distinguished the lightning rising from the ground, nor was its motion so swift but what we could discern its origin and direction."

Eight days out of Mexico City, the travelers arrived at Guadalajara and rested for two days before proceeding on to Mutchitilté. In the midst of mountains "piled up, as it were, one above another, which make it a most frightful situation," they saw a cascade that fell two hundred feet (sixty-one meters) in a sheet of water which "strikes the beholder with terror and admiration." There were other reasons for terror.

> It is impossible to conceive a more frightful and dangerous road than that which we travelled for near five leagues . . . this road, which is hardly four feet [1.2 meters] wide, is cut on the slope of a mountain that rises almost perpendicular; the road is about half way up, so that on one side you are hemmed in by the mountain, and on the other in danger of falling down such deep precipices, that in some places you hardly discern the tops of the tallest fir-trees in the vale below.

At one point they met a caravan of mules coming the opposite direction. Chappe watched, appalled, as the mules carrying his larger instruments walked a tightrope, as it were, on the very edge of the precipice.

On April 15 the expedition arrived at the coast and San Blas. Chappe learned to his chagrin that because of calms, contrary winds, and currents, the crossing to Cabo San Lucas could take as long as twenty-one days, even at a better season of the year. However, any thought of remaining on the near coast for the transit's observation was abandoned when he heard that the rainy season was fast approaching and would not abate until the end of June. Soon after they set sail on April 19, Chappe, in desperation and not knowing that he was giving a fatal order, told the captain that if they could reach the other shore, they should land at the first place possible—whether inhabited or not.

May 19: Their ship dropped anchor about a mile and a half from the coast, somewhat north of Cabo San Lucas. The captain was confident he could take them in safely, but bringing the instruments to shore would be risky in open longboats. Seeing water wash over the deck of the boat preceding his, which carried most of the instruments, Chappe feared most for his clock. He "wrapped it up very close, and sat down upon it myself, to keep it dry in case the waves should chance to wash us." Those who rowed the longboats seemed exceedingly skilled in the eyes of a worried Chappe.

> *The sailors on their part, attentive to the word of command, now rowed with all their might, now again stood stock still, either to avoid a wave ready to break over the boat, or to keep in the way of another that might wash us ashore. It was by this maneuver, executed with the utmost dexterity and success, that at last we got safe to land.*

That night Chappe lay by the waterside,

> *casting my eyes upon my instruments that lay all round me, and not one of them damaged in the least, revolving in my mind the vast extent of land and sea that I had so happily compassed, and chiefly reflecting that I had still time enough before me, fully to prepare for my intended observation, I felt such a torrent of joy and satisfaction, it is impossible to express.*

There was a Franciscan mission at San José del Cabo—a Jesuit community until two years before—about a mile from the beach. Chappe and his companions were welcomed warmly. Did no one realize that it was a welcome into the hands of death itself? What they called "an epidemical distemper"—probably typhoid fever—was raging at San José. One third of the population had died before Chappe's arrival, but he, convinced that the observation was more important than his own safety, chose not to move on to Cabo San Lucas.

Oblivious to the encroaching deadly threat, Chappe took over a large barn and removed the southern half of the roof, covering the opening with an awning in case of rain. He set up his instruments according to careful plans he had made during the voyage from Cádiz to Veracruz and attached his schedule to the barn's wall, where he could see it without leaving the instruments. The clock he'd protected so carefully in the journey to shore was a pendulum clock and, once set, needed stabilizing so that no jolt or other interference could mar its accuracy. Chappe had brought "a great beam of cedar" from the mainland, which he sank into the ground and braced against the walls of the barn and a brick stand. He attached the clock to this beam, in a box covered with paper to protect the mechanism from wind and dust. Finally, "All my instruments were fixed just as they were to stand to observe the transit of Venus. The weather favoured me to my utmost wish. I had full time to make accurate and repeated observations for the setting of my clock."

More and more of Chappe's hosts were dying, and Chappe himself could hardly have been unaware of the extreme danger he and his party were in, but, as Pauly would record, every day brought Chappe closer to the success of this enterprise and "Mr. Chappe cared for nothing else."

June 3: The day of the transit. Chappe's plan called for him to do the observing while a servant counted off minutes and seconds. Pauly, the engineer, recorded the times. Young Dubois tended the instruments. Just seconds before noon, Venus touched the disk of the sun, and this time Chappe did see the black drop effect. Six hours, twelve minutes, and fifty seconds later, the transit was over. Chappe made an ecstatic entry in his journal: "At last came the third of June, and I had an opportunity of making a most complete observation."

With those words, Chappe's journal ended. Perhaps he, only half-realizing what a short time he might have left, expected to take it up again.

It fell to his companions—for the most part, Pauly—to provide the rest of the story. At first it wasn't because of his own illness that Chappe failed to record journal entries. Two days after the transit, Doz and Medina fell ill. The scene of triumph had become a charnel house. "Every one, Indians, Spaniards, and Frenchmen, all were either dying or hastening towards death," Pauly wrote. He and Noël were the next to sicken. Chappe cared for people around him day and night but also continued to make observations. On June 6 he noted the time that Jupiter's moon Io disappeared behind the planet. He took observations of the sun to check the accuracy of his pendulum clock.

On Sunday, June 11, Chappe himself fell ill. Dosing himself with medicines he had brought with him, he improved enough to observe an eclipse of the moon on June 18 "with the utmost skill," Pauly recorded. Then he collapsed but nevertheless still managed to crawl to his instruments, observe eclipses of stars, and make solar observations. When his strength finally failed completely, he put the records of the transit and his other notes in a little box, hoping that someone would survive to take it back to Paris.

On August 1, Chappe's more than ten-year-long mission to observe and record the transits of Venus came to an end. He died surrounded by Pauly, Noël, and others, none with strength enough to crawl to one another or to attend Chappe's death. All local priests and missionaries had perished. Doz and Medina managed to perform the last rites. Chappe was buried as he had requested, in a Franciscan habit. Pauly, doubting that anyone would survive, took the precaution of asking a local native chief to put Chappe's box on a ship back to the mainland.

A ship arrived two months later. Medina, Doz, Pauly, and Noël were strong enough to board her, but Dubois, the watchmaker, was so near death that they left him behind. Soon after, horrified by what they had done, they went back and found him dead—probably, they feared, from despair at being abandoned. The four survivors were severely ill again by the end of the voyage to San Blas, and Medina died soon after landing. It fell to the severely weakened Pauly and Noël to undertake the long journey back across the continent and the Atlantic to bring Chappe's precious, dearly bought data to Paris.

In December 1770, Pauly delivered the box to the Astronomer Royale, who would write that Chappe's was one of the three most important and successful of the nearly one hundred fifty different observations of the 1769 transit. News of Chappe's death had preceded him. The Secretary of the Académie wrote his eulogy:

> *[Chappe d'Auteroche] had an open and candid, unpretentious soul, and a noble, straightforward and honest heart. . . . He was known in the highest circles; the King himself deigned to converse with him and honored his death with expressions of regret. Never was there one more unselfish than he. He liked fame, he wished to earn its favors, not to steal them. . . . One could only have wished that the last proof he gave, so worthy of praise, had not been fatal to him.*

The conclusions eventually drawn from the observations in 1761 and 1769 were not as definitive as Halley and the scientific world had hoped, partly due to the argumentative atmosphere in which the data was analyzed. Jérôme Lalande, who helped organize observations of both transits, calculated from his own data, Maximilian Hell's, and Chappe's, and arrived at results close to the modern Earth-to-sun measurement of 93 million miles, but Lalande's analysis became the target of criticism and controversy. The black drop effect and the apparent halo around Venus had again made it tricky to determine the instant the planet touched the sun's disk.

An illustration by Jean-Baptiste Le Prince from Voyage en Sibérie *entitled* La France et l'Empire, la Pologne et la Russie; *it depicts personifications of France, Poland, and Russia below a map of Chappe's travels in Europe.*

Edmond Halley had known that the distance to the sun would be a significant rung in the ladder to measure the universe. Journeying through Mexico, Chappe had pondered the importance of this hoped-for achievement and rejoiced that he was part of it: "If man, as an individual, is but a speck, an atom in this vast universe, he is, by his genius and his daring spirit, worthy to embrace its whole extent, and to penetrate into the wonders it contains." Many of the names and stories of the scholarly adventurers who observed the 1761 and 1769 transits of Venus are not entirely forgotten, and occasionally one is brought to light in a documentary or science magazine article, but history has not granted any of them the "immortal glory" Halley offered in his challenge.

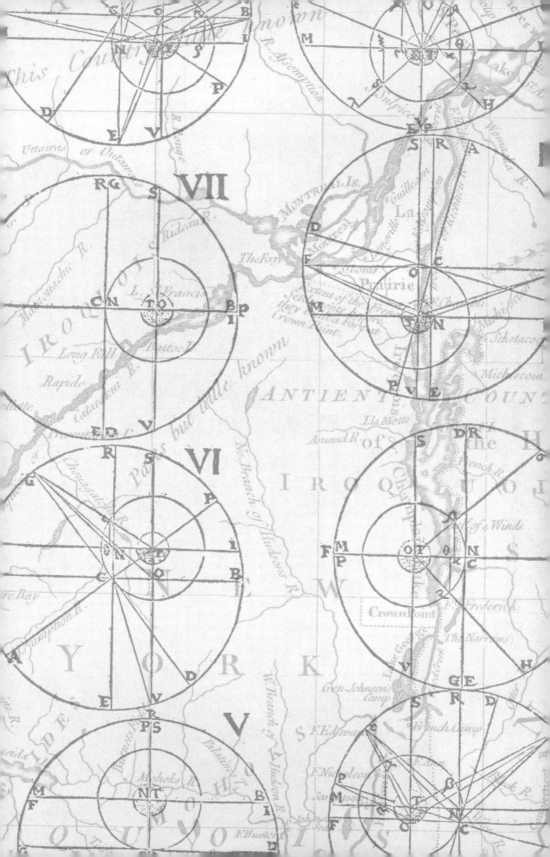

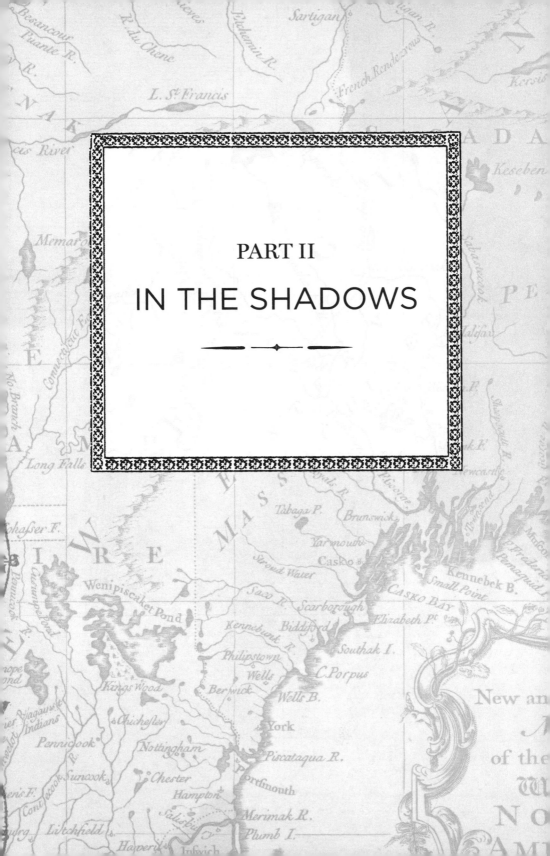

PART II

IN THE SHADOWS

FOUR

Mary the Jewess

LOST IN HER OWN LEGEND

(c. 1st–3rd centuries CE)

HERE ARE TANTALIZING EXAMPLES to be found in the early history of science and mathematics of people whose "real" life stories have been so lost in mazes of legends and distorted by centuries of retelling that we can't know for certain whether they ever lived at all. Authors, poets, playwrights, admirers, detractors, and even academic science historians have contributed to the fiction and the confusion. In the case of Pythagoras, for instance, it has become impossible to find out for sure who the man was behind all we think we know, or how much our popular image of him would be threatened by finding out his true story.

Mary the Jewess is someone who similarly might have become lost in her own legend—for the Mary the Jewess revered by medieval and early modern alchemists was largely a product of celebratory fiction. What little was known of her actual work was reinterpreted until it hardly resembled the real Mary's work at all, and the meaning she would have ascribed to it was distorted to suit a different age and culture than the one in which she lived.

Was there a *real* Mary the Jewess? We know that there was, and we know this because of one man. We see her through his eyes. Had he not written about her close to her own time and, more importantly, quoted her at length, and had not a good bit of his own writing survived, we would have nothing but the myth. His name was Zosimos, or Zosimos of Panopolis after his birthplace on the Nile a good distance south of Alexandria. He lived from the late-third century CE into the early fourth century.

OPPOSITE: *Illustration of Mary the Jewess from* Symbola Aureae Mensae Duodecim Nationum (Symbols of the Golden Table of Twelve Nations), *1617, by German alchemist Michael Maier.*

Zosimos spent most of his life at Alexandria, the center of the Hellenistic world. This splendid city, founded in the Nile Delta in 331 BCE by Alexander the Great, was an intellectual hothouse that encouraged enthusiastic study of the past and of earlier philosophers and scholars with a somewhat rigorous insistence on what could be considered reliable documentation. Zosimos himself was a man who collected knowledge and a prolific scholar. He and his sister (or "mystical sister") Theosebia

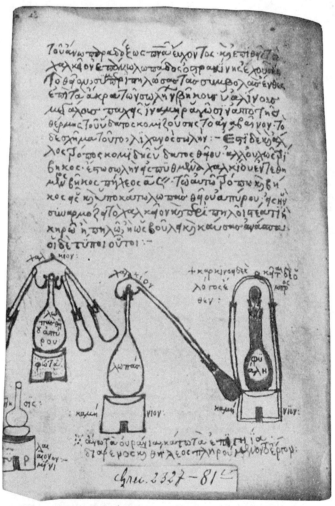

A fifteenth-century Byzantine Greek manuscript page illustrating Zosimos's distillation equipment.

compiled a twenty-eight-volume chemical encyclopedia, and he wrote at least twenty-two treatises as well. Throughout his work, Zosimos quoted at great length from earlier alchemists, giving a window into alchemy in Alexandria prior to his own time, and he focused particularly on two people: Mary the Jewess, also called Maria Hebraea or Mary the Hebrew (or Maria Prophetissima by later alchemists); and a man who has come to be known as Pseudo-Democritus—"pseudo" so as not to confuse him with the philosopher Democritus who lived some five hundred years earlier.

In the first centuries CE, Alexandria was already ancient, and its large, prosperous Jewish community—of which Mary the Jewess was a part—was as old as the city itself. Through the years, its members had worked carefully to stay on good terms with their Greek and Egyptian neighbors and later with the Romans who ruled Egypt after 30 BCE. By and large, their efforts succeeded. Except for periods of persecution during the reigns of the Roman emperors Caligula, Nero, and Vespasian, the Alexandrian Jews were treated with respect and were free to practice their religion. It was within this community that Mary the Jewess grew up—never mind that some writers and legends placed her at least sixteen centuries earlier and in other parts of the world. She was a real person, but she lived at what British science historian Jacob Bronowski called "the hinge of legend and history," where the lines between fiction and fact are difficult to discern. Later Hellenistic and medieval alchemists thought of her as the founder of their art and the ultimate authority when it came to both theory and practice. Her aphorisms were, to their minds, prophetic, and alchemists revered her as a prophetess even into the late sixteenth century.

Though there are stories about early alchemists that make claims for much more ancient origins, the first reliable scholarly sources about alchemy in the Western world date from no earlier than about 300 CE, and these sources could only trace its history with certainty as far back as women and men who lived and worked in Alexandria in the three hundred years prior

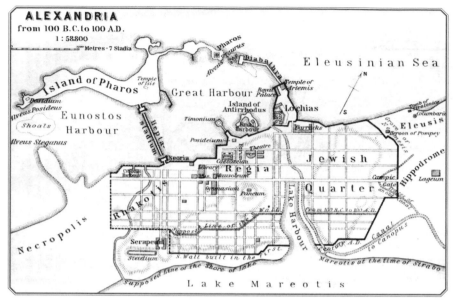

A 1914 map of Alexandria as it looked when Mary the Jewess lived there.

to that. In this scholarship, Mary the Jewess was reputed to be the earliest alchemist, though, from reading what is known about her and what she accomplished, few would claim that she was doing it all completely from scratch. There are no surviving alchemical treatises earlier than Zosimos, from whom the most trustworthy information about her comes. However, he was definitely not Mary's contemporary. He called himself her disciple, but that should not be taken to mean he actually sat at her feet. Rather, he followed—and adhered to—her way of thinking and working.

The alchemists of Hellenistic Egypt from the earliest we know about them held the belief that their knowledge stemmed from secret, sacred Hebrew alchemical science, known only in its purest form to Jews and supported by a biblical prehistory. Zosimos seems, judging from quotations attributed to him, not to have shared this belief, but he did think of the Jews, and particularly of Mary, as the most important sources available to him of knowledge that Egyptian and Greek alchemists had kept secret and not passed on to his and succeeding generations. Zosimos was not a Jew—he was a Gnostic Christian—but he readily assigned a lofty status among alchemists to the Jews. He quoted a passage from Pseudo-Democritus:

> *It was the law of the Egyptians that nobody must divulge these things in writing. . . . The Jews alone have attained knowledge of its practice, and have described and exposed these things in a secret language.*

Zosimos wrote about the alchemists in the third person, as though he were an outsider, while at the same time making it clear that if he *had* been an outsider, he would not have been privy to this information. According to him, the monopoly held by established members of the alchemical community (who were not necessarily all Jewish) was upheld and protected by the government of the city.

> *He who made a search without authorization could be thrown down [and put to death] by the overseers of the town markets. . . . Likewise, it was not allowed to operate the furnaces secretly, or to fabricate in secret proper tinctures. Also you will not find anybody among the ancients who reveals that which is hidden, and who exposes anything clearly in this regard.*

Later alchemists would connect many biblical characters—particularly Adam, Moses, and Solomon—with alchemy despite little or no support from the biblical text or any other historical evidence and come to believe that their art was a privileged, secret, divinely inspired science that stretched back to creation. Though Mary the Jewess—according to quotations attributed to her—wished to limit the most secret knowledge of the "philosopher's stone" to "the race of Abraham," other ancient scholars traced a history of alchemy that included non-Jewish alchemists in Persia and Egypt.

When was Mary born? How long did she live? Was Alexandria a peaceful city in her time, or was she alive during one of the Roman persecutions of the Alexandrian Jews? Perhaps she died in one of them? What of her family? Did

A three-headed monster in an alchemical flask representing the composition of the alchemical philosopher's stone: salt, sulphur, and mercury; the illustration is from a sixteenth-century alchemical work titled Splendor Solis *(Splendor of the Sun).*

she remain celibate, as some scholars of both sexes chose to do in northern Egypt during her time? Such questions are unanswerable. We don't even know how far removed in time Zosimos actually was from Mary. Judging from his writings, he was unwilling to commit to dates concerning her life and death, even though his sister Theosebia specifically asked him for that information.

Scholars have tried to put some of the clues together: Zosimos refers to Mary as one of "the ancients," "the sages," or the "first of the ancient authors" and calls a treatise probably written by her—"On Furnaces and Apparatuses"—a "writing of the ancients." "First of the ancient authors" is probably the most telling phrase here. In order for her to have been the first—considering others Zosimos mentions who lived before him but after her—she had to have flourished at least two generations earlier than he did. Probably she was earlier than that, for a mere two-generation gap would seem to make it possible that he knew her or knew about her on a closer basis than his writings indicate. A long enough time had to have passed between her lifetime and Zosimos's for him to indicate without hesitation that she played a seminal role in the development of alchemy in the Hellenistic world. On the other hand, she can't have lived earlier than the second half of the first century BCE, for that was when neo-Pythagorean and neo-Platonic philosophy, with which Mary was familiar, arrived in Alexandria with the appearance of the philosopher Eudorus. From clues such as these, scholars have tentatively assigned her to sometime within the first to early third centuries CE.

A drawing recreating what the cultural heart of Alexandria may have looked like in the fourth century—a century or so after the age of Mary the Jewess—by nineteenth-century English painter Robert Trewick Bone. It depicts the Greek philosopher Hypatia teaching at Alexandria.

We do know something about the community in which Mary grew up and later worked. Jewish homes and synagogues were located all over the city of Alexandria during those centuries, but Mary very likely spent her childhood in a large Jewish section near the royal palace, not far from the harbor (considered an advantageous location in those days of sea-going trade). This district had been given to the Jews by the first rulers in the Ptolemaic dynasty, ostensibly so that they could practice their religion without being hindered by contact with their pagan neighbors. We know nothing about Mary's family or what roles they may have played in this community, but they would have been part of a political entity with its own laws and judges, independent of the council and civil government of the rest of the city, though its members remained full citizens of Alexandria and sometimes held high civil and military offices.

Though Rome was the center of military power, Alexandria, with its wealth, climate, mix of cultures, and its splendid library, begun by the Ptolemy shortly after the founding of the city and unrivaled in the ancient world, had become—as Athens declined—the intellectual capital of the

An illustration of Philo of Alexandria from Les vrais pourtraits et vies des hommes illustres . . . *(The True Portraits and Lives of Illustrious . . . Men), 1584, by French Franciscan explorer and cosmographer André Thévet.*

world. The Jews who lived there were indeed thrice blessed. They had the advantage of a superbly impressive ancient tradition of their own. They lived in a city rich in the learning, philosophy, and pursuit of knowledge inherited from the Athenian Greeks. The mystery of ancient Egypt was at their doorstep. Historian Raymond F. Surburg described Philo of Alexandria—the great Alexandrian Jewish philosopher of the time of Christ—with the words: "By nature and upbringing he was a Jew; by residence in Alexandria a mystic; by higher education a Greek humanist; by contact and social position an ally of the Roman aristocracy." Surburg wrote that Philo represented "a strange fusion," but it was not so strange or unusual in Alexandria. It was in this remarkable context that Mary the Jewess became a highly respected scholar.

Zosimos had no problem with the idea of a woman being one of the most knowledgeable of the ancients. In one fragment attributed to him, he passed along a legend that women had learned the art of metallurgy from fallen angels whom they had married. This lore, of course, robbed the women of any credit they deserved:

> *The ancient and divine writings say that the angels became enamoured of women; and, descending, taught them all the works of nature. From them, therefore, is the first tradition, **chema**, concerning these arts; for they called this book **chema** and hence the science of chemistry takes its name.*

This involvement of women with alchemy—if indeed there *were* others among Mary's contemporaries—seems not to have continued. After her, almost all alchemists were men until about 1700, the one exception being a Jewish woman, whose name is unknown, whom later manuscripts described as "expert in the work."

Many chemists today are dubious about the idea that their field is an offspring of alchemy, yet it is not without good reason that Mary the Jewess has been called their ancestor. To judge from Zosimos's quotations from her, she did indeed either birth or pass on faithfully a vocation whose practitioners design, redesign, utilize, record, and evaluate the results from chemical experiments using specialized laboratory equipment. Thanks to Zosimos we know a great deal about the pieces of equipment Mary used. She probably invented these, for the first mentions of them are in her words, quoted by Zosimos, in which she tells how to construct them and describes them in detail. Mary's most significant and long-lasting contributions were basic, practical ovens and apparatuses for cooking and distilling. Their descendants are still used in modern chemistry. Mary constructed her equipment from metal, clay, and glass, the latter of which she favored because glass allows one to "see without touching." This was essential when working with mercury, which she described as "deadly poison, since it dissolves gold, and the most injurious of metals," and "sulphurous [sic]," a word for substances containing arsenic. She probably did not blow the glass used in her equipment herself, so she must have worked closely with glassmakers to get exactly the right result. To tighten and seal connections in these instruments, she used fat, wax, starch paste, fatty clay, and something called the "clay of the philosophers."

Among Mary's laboratory equipment, the piece that would enjoy the most longevity was the *balneum Mariae*. It was essentially a double boiler consisting of an outer vessel filled with water that could be heated and that would in turn slowly heat something in an inner vessel, holding the contents at a constant temperature without boiling it. Two thousand years later, in modern France and French cooking, the double boiler is still called a bain-marie—Mary's bath.

Hers also is the oldest recorded description of a still. As stills commonly do, Mary's had three parts: There was a copper vessel with a heat source

A bain-marie, or "Mary's bath," distillation apparatus, illustrated in Coelum philosophorum . . . (The Heaven of the Philosophers), *1528, by German professor Philipp Ulstad.*

to heat the material being distilled, a cooler "still head" where the steam from the heated material would condense, and lastly a receiver vessel or vessels. Mary's still, called a *tribikos*, had three receiver flasks and three copper delivery spouts leading from the still head to the flasks, possibly with a sort of gutter inside the still head to collect and transport the distillate to the spouts. From her description, it seems that one of the spouts collected distillate from higher in the still head than the other two. She gave instructions for creating the copper tubes from sheet metal, which should be "a little thicker than that of a pastry cook's copper frying-pan." She recommended a paste of flour to seal the joints. (See Figure 1.)

One of Mary's main interests was in the action of the vapors of arsenic, mercury, and sulfur on metals. For this purpose she developed the *kerotakis*. The name and idea came from a triangular palette that artists used to keep mixtures of wax and color pigments hot. In Mary's adaptation, metals, rather than wax, were softened. Her apparatus consisted of a sphere or cylinder with a domed cover. She set this over a fire. In a pan near the bottom, she heated up solutions of sulfur, mercury, or arsenic sulfide. She suspended copper-lead alloy on a palate near the top of the cylinder. As the sulfur, mercury, or arsenic sulfide boiled, the steam condensed at the top of the cylinder in the domed lid and then ran back down the sides of the cylinder in a continuous flow. The vapors acted on the metal alloy on the palate, making a black substance that

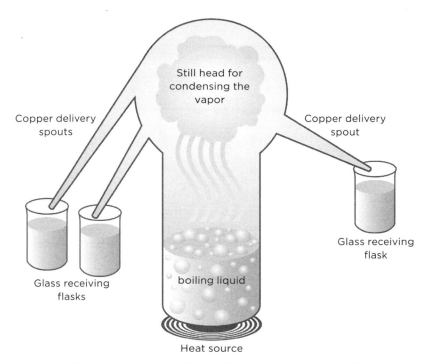

Figure 1. *An illustration of Mary's* tribikos *still, which had three receiver flasks and three copper delivery spouts leading from the still head to the flasks.*

came to be known in alchemy as Mary's Black. This, according to alchemists, was the first stage of "transmutation," the conversion of base metals into precious metals. Continued heating in Mary's *kerotakis* ended not in gold but in a gold-like metal alloy. Mary also used her *kerotakis* for the extraction of attar of roses and other plant oils. (See Figure 2.)

With the invention or adaptation of these apparatuses, her meticulous descriptions and instructions, and the information and theories she extrapolated from their use, Mary established many of the theoretical and practical underpinnings of Western medieval alchemy and, via that route, of the science of chemistry. She stressed the concept of the macrocosm and the microcosm, looking for the same patterns repeated in all levels of the cosmos from the largest scale to the smallest. Though she clothed it in rather metaphorical language, there was nothing particularly unscientific about

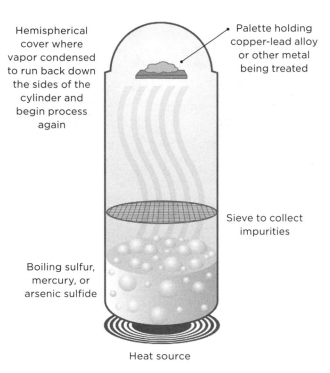

Hemispherical cover where vapor condensed to run back down the sides of the cylinder and begin process again

Palette holding copper-lead alloy or other metal being treated

Sieve to collect impurities

Boiling sulfur, mercury, or arsenic sulfide

Heat source

Figure 2. *A simple diagram illustrating Mary's domed apparatus, the* kerotakis. *A pan near the bottom heats solutions of sulfur, mercury, or arsenic sulfide, while copper-lead alloy is suspended on a palate near the top. As the sulfur, mercury, or arsenic sulfide boil, steam condenses in the domed lid and runs back down the sides. The vapors act on the metal alloy on the palate, "transmuting" them.*

her hypothesizing that metals were or resembled living beings, male and female. In her laboratory experiments, she found that some metals "couple" more readily than others. Silver, she said, coupled easily, while copper coupled about as well "as the horse with the ass, and the dog with the wolf."

The lengthy quotations from Mary in Zosimos's treatises show that she was a meticulous experimenter. She gives exact measurements and timings ("Leave it for three times twenty four hours"), discusses ingredients and teaches how and how not to use them ("Remove the sulphurous from the lead, [for] wherever the sulphur enters, it will tint"), warns about

their dangers ("Don't touch it with your hands, for this is an igneous preparation" and "One must beware of stirring the mixture with the hands, for
the mercury is deadly"), indicates colors and tones of colors to be watched
for ("When it loses its glitter, one combines it with the gum"), tells how to
apply heat and fire ("Make it digest in the heat of manure"), indicates the
results to be expected ("Copper, refined by melting, diminishes by one-
third of its weight"), refers readers to other parts of her writings, and gives
them reader-friendly encouragement ("Do not be afraid to try in detail all
the things which I have revealed to you" and "Do not be anxious to know
whether the work is on fire"). So she continues page after page.

Mary also lays out in careful order the progression in which things
must be done. Here, for example, are Mary's instructions for one of her
simplest processes, quoted by Zosimos:

> Copper burnt with sulphur, treated with oil of natron*, and recov-
> ered after having undergone the same treatment several times,
> becomes excellent gold, and without shadow. . . . [if you start] by
> burning the copper, the sulphur produces no effect. But if you
> [start by burning] the sulphur, then it not only renders the copper
> without blemish, but also makes it approximate gold.

One of the primary goals of alchemy, a dream that made it a ruinous,
addictive obsession for many of its practitioners through the ages, was the
transmutation of base metals into gold and silver. Judging from Mary's
writing, she never expected to create gold, only something that looked like
gold—or would "approximate gold."

The other goal of alchemy was to concoct a potion, an "elixir of life,"
that would give the drinker eternal life or eternal youth. None of the quotations that Zosimos had from Mary indicate that she interested herself in
that effort. In his quotations, "divine water" should not be confused with
the "elixir of life." According to Zosimos, "divine water" was liquid made

---

* Natron is a salt-like substance that the ancient Egyptians gathered from dry lake beds. They used it,
blended with oil, as we use soap, and also used it in the mummification process.

with two parts of lime and one part of sulfur, boiled, filtered, and boiled again. The words for "sulfur" and "divine" were very similar in pronunciation, which may explain its name. Quoting Mary,

> *This divine water, whitened by the whitening materials, whitens.*
> *Yellowed by the yellowing materials, it makes yellow. Blackened*
> *by means of vitriol and gallnut, it blackens, and carries out the*
> *blackening of silver and of our molybdochalkon [alloy of cop-*
> *per and lead]. . . . Thus the water blackens, attaching itself to our*
> *molybdochalkon, and gives it a permanent black tint.*

She goes on to tease her readers, "Even though this tincture is unimportant, all [you beginners] desire very much to know it." She also used "divine water" for coloring stones, and Zosimos paraphrased her instuctions in his own treatise "The Coloring of Precious Stones," saying that this is "according to the precepts of the Hebrews."

There was another side to Mary's writing that was not so down-to-earth. After all the practical advice and applications, it comes as a surprise to find parts of her teaching—again as conveyed through Zosimos—couched in the mystical language that we associate with medieval alchemy. Though writers of many centuries have readily stressed this link, most have failed to look for an explanation of Mary's usage and preference for this kind of language and thinking in the time in which she lived. What is most interesting is how much her aphorisms sound like neo-Pythagorean and neo-Platonic philosophy of the final century BCE and the first centuries CE. These philosophers were not alchemists. For example, Mary's teaching that "One is All and All is One" is easily passed off as typical alchemical jargon. It became that in the Middle Ages, but in the intellectual milieu of ancient Alexandria it was a connection with cutting-edge philosophy.

A little background is in order: Among the neo-Pythagoreans and the neo-Platonists—the two are difficult to distinguish because neo-Platonists assumed that Plato got his philosophy from Pythagoras—there was a developing trend toward the idea of one supreme transcendent god.

Earlier philosophers had devised "tables of opposites" and argued about what should appear where in the table. Should "One" be at the top? Did the One include the All, or was it opposite to All (plurality)? Plato had written about the "One." So had Aristotle, and on a table recorded by him, One and Plurality are listed as opposites but are not at the top of the table. In the second half of the first century BCE, Eudorus of Alexandria—considered to be the first to bring neo-Pythagorean ideas to that city—argued that in Pythagorean doctrine, the One transcended everything else. His table of opposites put the One at the top, astride both columns and without an opposite. This was not actually monotheism but was leaning in that direction.

There is no reason to assume that Mary was speaking in terms of alchemical formulas or mystical incantations when she declared that All and One were the same thing. These were philosophical concepts with which any Alexandrian intellectual would have been familiar. Mary was also coming to the discussion not from a pagan direction, but as a Jewess who always referred to "God" in the singular. The hint of pantheism in her aphorism "One is All and All is One" made her very much an Alexandrian Jew of her time. Though many of her neighbors in the Jewish community were remaining faithful in attendance at Jewish services and keeping the Jewish laws, their thinking and personal faith had assimilated much of the intellectual atmosphere of a very cosmopolitan city.

Mary was reported, not by Zosimos but by a man he calls Christianos, not merely to have written but to have "shrieked" another aphorism: "One becomes two, two becomes three, and by means of the third and fourth achieves unity; thus two are but one." That is a lengthy shriek, and we can hope it didn't shatter any of her glass vessels. It might sound to modern ears like a magical incantation, perhaps from a Harry Potter novel. But in what-ever tone of voice Mary may have uttered these words, however many times they were repeated as a mystical aphorism by later alchemists, and however much some modern writers believe this was only a formula for an alchemi-cal procedure, they are actually part and parcel of something much earlier: again, Pythagorean Platonic philosophy.

Greek philosopher-mathematician Pythagoras, shown in an engraving from The History of Philosophy . . . , *c. 1655, by English author Thomas Stanley.*

Though Plato almost certainly had not derived all his philosophy from the Pythagoreans, he had carried forward the great Pythagorean theme: the underlying mathematical structure of the world and the power of numbers for unlocking its secrets. Mary would certainly not have been immune to those considerations and discussions. Plato and other great minds had been scrambling to get a foothold on the climb to dependable answers about the world and its creation. If mathematics and numbers were the clue, how should they be used? The numbers one, two, three, and four, about which Mary may have shrieked, were the numbers in the Pythagorean *tetractus* (see Figure 3)—an equilateral triangle that incorporated the ratios that the Pythagoreans had discovered in musical harmony—making those numbers seem a particularly good starting point for any investigation of nature or its source. They were the most significant numbers in a creation scheme that Plato had picked up from the work of the fifth-century Pythagorean philosopher Philolaus. There were certainly many others besides Mary who thought these were significant foundational numbers. There could have been a lot of shrieking going on.

To Philolaus the numbers one to ten were all important, and ten was, to Pythagoreans, the "perfect number," but one, two, three, and four were the numbers that added up to *make* ten (illustrated in the *tetractus*) and that

Figure 3. *The Pythagorean equilateral triangle known as the* tetractus. *The numbers one, two, three, and four (as shown in the rows of the tetractus) add up to the "perfect" Pythagorean number: ten.*

produced geometry, giving us the three-dimensional world we experience. As much as we think we know better than Philolaus, we can appreciate the line of thinking in his creation scheme: a point represents the number one; a line is two (defined by two points, one at each end); a triangular surface is three (defined by three points, one at each corner; and also made of three lines); a *solid*—a pyramid—is four (defined by four points, one at each corner, and also made of four triangles). Thus two (the line) and three (the triangle) achieve unity in four (the pyramid). That pyramid represents the three-dimensional world of Philolaus's and our own experience. Hence the pyramid can be considered unity, or "One," and become a point from which to start the sequence over again on a larger scale. And there it is: Mary's "One becomes two, two becomes three, and by means of the third and fourth achieves unity." (Plato's own creation scheme, described in his *Timaeus,* used the numbers one, two, three, four [the square of two], nine [the square of three], eight [the cube of two] and twenty-seven [the cube of three].) Clearly there was a lot of thought and manipulation going on having to do with the four numbers about which Mary shrieked. Nor did this preoccupation end with the ancients. The nineteenth- to twentieth-century philosopher Carl Jung believed there is deep psychological meaning in the transformations of one into two, and two into three, and the way one emerges as the fourth, out of the third.

For a scholar in ancient Alexandria, a connection with high-minded, abstract philosophical discourse did not rule out having one's name also celebrated for mysterious skills with magic and the occult. Among a great many people in Mary's time, Pythagoras was revered not for any ideas about numbers but as a semi-divine personage, a powerful magician, a shaman who had practiced secret, hidden arts. To be fair, his followers *had* been intensely secretive. But it was not until later generations after Mary that alchemy became almost synonymous with occult practices, involving incantations and what to most modern ears sounds like complete nonsense. Aphorisms like Mary's—disassociated from their philosophical intellectual origins—provided much grist for that mill, and medieval alchemists gave Mary's sayings a much more mystical, magical spin. Meanwhile, in the hands of others—Tycho Brahe, for example—alchemy rooted in Mary's principles was still exacting, methodical, and practical. It was alchemy that allowed Tycho to fashion a false nose for himself out of gold and silver when his own was sliced off in a duel. It was also an addiction to alchemy that destroyed his brother-in-law, Erik Lange, and impoverished his sister, Sophie Brahe. There were indeed two sides to medieval and early modern alchemy, but the most clear-minded, careful experimentation evolved into the modern science of chemistry.

Meanwhile, in the years after Zosimos, Mary the Jewess's name and work became the stuff of legend. She began to be revered as one of the ancient seers. There were more specific stories: One fourth-century Christian bishop, Epiphanius of Salamis, reported angrily in his book *Against Eighty Heresies* that Mary had claimed to have a vision of Christ. This made her someone to be either reviled or revered, and the divide did not necessarily follow Christian/non-Christian lines. Epiphanius condemned two works supposedly authored by her—*Great Questions* and *Small Questions*—which, if his description of them had been accurate, lent a rather seamy twist to some of the sayings of Jesus, either in the form of erotic metaphor or pornography. It is no wonder the bishop was perturbed. However, there is no evidence whatsoever that any such books ever existed nor any trace of their names or subject matter in any other writing. No other author in the fifteen hundred

Miriam, sister of Moses, dancing, 1864, by English artist Edward John Poynter. Some early legends mistakenly equated Mary the Jewess with Miriam of the Old Testament.

years after Epiphanius—a time when Maria remained a key figure in alchemical lore—ever referred to this particular link with Christianity.

Other legends appeared and were repeated for centuries. An Arabic source called her "Maria the Copt" (which means "Mary the Egyptian"). There were Coptic Christians, and this source pictured her carrying the infant Jesus on her shoulder while holding a spindle in one hand. Another popular story placed her back five centuries before Christ and made her a student of the celebrated Persian alchemist Ostanes, who was not a fictional character.* He lived from circa 519 to 465 BCE and was the brother-in-law of Xerxes I, the son of Darius the Great. One tale had Ostanes not in Persia but teaching at Memphis in Egypt, where Mary was supposedly either his fellow instructor or his disciple. A later medieval Arabic book calls Mary "Mariya the Sage, daughter of the king of Saba." Here she is not a follower of Christ but of Allah, and her supposed words about the philosopher's stone were reported as "This is a great mystery. It is scorned, and trodden underfoot. But this scorn is a grace from Allah, may He be exalted, so that the fools should not know it, and it be forgotten." In the thirteenth century, Arnaldus de Villa Nova

---

* Ostanes was possibly multiple people with the same name.

said that Mary was the "daughter of Pluto," the Roman god of the underworld. Elsewhere she was Miriam, the Old Testament sister of Moses and Aaron, which would have dated her approximately sixteen centuries before the years she actually lived.

In the seventeenth century, chemistry emerged as an experimental laboratory science. Robert Boyle, one of its founders, finally brought an end forever to the belief in four basic elements—earth, fire, air, and water—which had gone virtually unquestioned for more than two thousand years (except for the sporadic addition of a fifth, *aether*) by everyone in the intellectual Arabic and Western worlds, including Mary the Jewess. In this same century, Johannes Kepler published his *Astronomia Nova*, Galileo used the telescope, and Isaac Newton wrote his *Principia*. The age of "modern science" had begun. However, especially early in the century, many influential people including the Holy Roman Emperor Rudolf II considered alchemy and astrology to be as respectable and important as astronomy. A physician, alchemist, and musician named Michael Maier, who worked in Rudolf's court in Prague at the same time as Johannes Kepler, wrote musical canons to serve as a "technical aid" when working with mystical formulas. Maier was also still popularizing stories about Mary the Jewess, repeating another long-accepted legend that "Maria the Hebrew" was "successor" to Hermes Trismegistus—a sage and sorcerer who, in Hellenistic and Roman–Egyptian mythology, was popularly believed to have lived near the time of Moses. Hermes Trismegistus, it was said, had authored many books on magic and the supernatural, alchemy, astrology, theology, science, and philosophy.

The Pythagoreans, along with Plato, Aristotle, and Mary the Jewess, were not wrong about a mathematical structure and rational order underlying the confusion and complexity of nature, but they had no conception of how deep and hidden that structure usually lies. It is no wonder that some of their attempts to plumb those depths sound ridiculous to us today. The Pythagoreans had happened to discover the rationality of the universe in one of the few places where it is close to the surface of things—musical harmony. But for centuries, people would go on looking for this structure on far too superficial a level. Even Kepler's third law of planetary motion was an almost accidental discovery that he made while trying to study and prove a much more superficial and, by modern standards, off-the-wall theory.

Mary the Jewess, however much interest she may have had in the philosophical discourse of her day, chose mainly to pursue her investigations on a much more practical level, with her handy inventions, her clear instructions, and her meticulous daily experiments. Though few would place great value on the work of alchemists who both revered her and thought they were following her example in the Middle Ages and early modern world, they did preserve something of her tradition for future generations of modern chemists. Thanks to Zosimos, the woman she really was is not completely lost to the history of science.

FIVE

Maria Sibylla Merian

WONDROUS TRANSFORMATIONS

(1647–1717)

HE YEAR WAS 1700. JUNGLE UNDERGROWTH clung to the woman's skirts like wet fingers as she squinted at the canopy high above. She wasn't young but in her fifties, a European far from home, in the Dutch colony of Surinam (present-day Suriname, South America) . . . and currently frustrated because she couldn't investigate the treetops. But she wasn't always defeated. On one occasion her hired servants climbed a palm tree and brought down a large web of caterpillars. They carried the whole sticky mess back to her house in Paramaribo so that she could study the insects, take notes, and capture one in a drawing. Her helpers sometimes cut down whole trees to enable her to examine the creatures in their leaves.

MARIA SIBILLA MERIAN
Nat: XII. Apr: M D C X L V II . Obiit XIII. Jan. M D C C X V II .

Not everything Maria Sibylla Merian found was beautiful, and she wasn't content just to seek out attractive caterpillars, butterflies, and plants and draw lovely pictures of them, though she was better capable of doing that

OPPOSITE: *Plate XI from Maria Sibylla Merian's* Metamorphosis insectorum Surinamensium, *1705; the plant is a swamp immortelle* (Erythrina fusca Loureiro); *the insect moth and larva is a giant silk moth* (Arsenura armida). LEFT: *An eighteenth-century Dutch engraving of Maria Sibylla Merian.*

115

than any other artist of her day. Her passion was studying and recording the lifetimes, life cycles, and lifestyles of insects—which caterpillar changed into which pupa, which pupa into which butterfly or moth, to feed and lay its eggs on what plant. And what might happen to it along the way. Why did some pupae erupt in nasty flies instead of moths? What ate what? She put all of it into her drawings. In the introduction to her book *Metamorphosis insectorum Surinamensium*, she wrote that she "described them from life and placed them on the plants, flowers and fruit on which they were found."

It had been relatively easy to satisfy her curiosity in her garden and the countryside where she had lived in Frankfurt, Nuremberg, and Amsterdam, and even in the restrictive religious community she joined for a time in Wiewert. She had collected many live specimens so that she could watch them develop and change. But in Surinam, too much of what she longed to study was out of reach. In the jungle canopy there were butterflies that she would never see, and even if one of them *were* to venture into reach or fall from the crown of the trees, it was impossible to discover what caterpillar it had come from or on what leaves it laid its eggs. She left Surinam sooner than she had planned, ill with either malaria or yellow fever, discouraged that she had not been able to accomplish all she'd hoped, but nevertheless able to produce a splendid book of drawings and descriptions. No one before her had chronicled metamorphosis so thoroughly and in such detail, nor shown the connections that link a butterfly—she called it a *Sommervögel* (summer bird)—or caterpillar with its life history and environment.

Who was this woman, so obsessed with these creatures that all her life she kept an eye out for a specimen here, an example there, a new kind here, a mysterious appearance there? Who was she, able to put what she found down on paper, in words sometimes but mainly in magnificent pictures? And why, today, does she tend to be remembered, when she is remembered at all, for her art . . . not for her science?

When it came to skillful drawing and engraving, Maria Merian had been given a head start. Her father's publishing house and printing shop in Frankfurt produced large, exquisitely engraved maps, colorful drawings and

Maria Sibylla Merian's father, publisher Matthäus Merian, in an engraving, c. 1630.

prints, and expensive books. Matthäus Merian had printed editions of *Les Grands Voyages*—initially the work of his first father-in-law, Théodore de Bry—which told stories of journeys to the New World. The workshop would also have had frequent visits from men working in science, other intellectuals, and more unorthodox and sometimes revolutionary characters, all interested in having their ideas correctly recorded and published in printed books. As Maria Merian's biographer Kim Todd described in his book *Chrysalis*, such an establishment was like "a coffeehouse before anyone in Europe drank coffee, where the heady brew was ink."

This busy, stimulating environment was the scene of Maria's early childhood. As she grew older, she would have taken her own place in that workshop, at least on the level of an apprentice and probably helping run the business, as women in families like Matthäus's often did. But Matthäus died when Maria was only three years old; and when her mother remarried a year later, Maria found herself in a new kind of household with the possibility of a different sort of education. Her stepfather was Jacob Marrel, an art dealer and painter, a pupil of the artist Georg Flegel, whose drawings of plants and insects, particularly beetles and wasps, were arguably the most convincingly naturalistic before Maria's. Marrel preferred to paint still lives of flowers.

Detail of a map of Frankfurt, the Merians' hometown, printed by Matthäus Merian, c. 1630.

Guild rules in the late 1600s didn't allow women to work with oil paints, but they could use watercolors. Maria first learned by copying. Her earliest surviving sketches and watercolors are of a grasshopper copied from Marrel and a Narcissus copied from Théodore de Bry. She used chalk to draw a grid and then used the grid to keep her work accurate and in scale. Engraving was another skill that she acquired, possibly from her half-brothers at the printing workshop. Maria also learned to read and write, perhaps in school. Frankfurt required—though did not always enforce—elementary schooling for both girls and boys.

Soon Marrel was sending her on small expeditions to collect insects to use as models for his own drawings, and he encouraged her, when she drew flowers herself, to include caterpillars, butterflies, and "such little animals, as the landscape painters do. They make one alive through the other." However, it wasn't until Marrel left the family to return to the Netherlands that Maria, at age thirteen, began in earnest the work that would engage her for the rest of her life. She recorded the date in a later *Studienbuch*. After describing the way she raised and hatched cocoons, she wrote, "This research I started in Frankfurt in 1660. Thank God." She also mentioned these early beginnings many years later in her book *Metamorphosis insectorum Surinamensium:*

> [I collected] all the caterpillars that I could find, in order to observe their metamorphosis. I therefore withdrew from society and devoted myself to these investigations; at the same time I wished to become proficient in the skill of painting in order to paint and describe them from life.

Maria grew up at a time when it was widely believed that insects appeared by "spontaneous generation" in putrid, rotting meat and vegetables and other garbage, in dead animal or human bodies left too long, and in mud and moist places. No eggs. No parents. No ancestors. Serious efforts to classify insects were a thing of the future, though some were grouped by what they ate; in other words, mostly by what they destroyed. Not everyone, however, thought that insects originated in spontaneous generation, and few believed that all did. The process of metamorphosis had been known in the East since 5000 to 3000 BCE. In China, breeding silkworms to produce silk fabric had for many centuries required expert knowledge of the different stages of this creature's lifetime, as well as how to protect and nurture it at each stage.

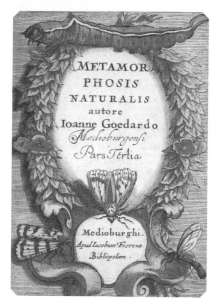

Title page of Metamorphosis naturalis, *1662, by Johannes Goedaert, a book that may have inspired Maria in her own work.*

When she was in her early teens, Maria encountered a book, the first volume of three, titled *Metamorphosis Naturalis* by Dutch painter Johannes Goedaert, which appeared between 1662 and 1667. The book resembled those she would later create herself in that the plates show insect metamorphosis in carefully observed detail. However, Goedart's work differed from hers dramatically in that the pictures of larvae, pupae, and adult insects are static drawings and don't show the plants on which the caterpillars fed or where the pupae could be found. Goedaert had not entirely abandoned the idea of spontaneous generation, but he understood that silkworms were not the only caterpillars that changed into moths or butterflies. His book may have been Maria's inspiration as she began capturing and collecting specimens, breeding them in her room, and recording their progress in precise drawings.

One of Maria's stepfather's apprentices in Frankfurt had been a young man named Johann Graff. Graff returned in 1664, when he was almost thirty years old and Maria was sixteen. In 1665 they married. Three years later, their first daughter was born. They named her Johanna.

By that time, those natural philosophers who accepted metamorphosis as the way most insects developed were disagreeing about something else: how much of the final structure of the butterfly or moth was actually present, contained and hidden somehow, in the caterpillar and pupa? Was the entire adult creature already there in miniature? Maria experimented with pupae herself and wrote:

When put on a warm hand it started moving vividly and you could clearly see that inside the changing caterpillar, or better inside its date kernel [her name for pupa], was life nevertheless. But if cut open too early or after two days, nothing but colored watery material comes flowing out.

In 1670, Graff, Maria, and baby Johanna moved to Nuremberg, Graff's birthplace. Maria was now calling herself Frau Gräffin. Her work with insects was beginning to be noticed and respected. Joachim von Sandrart, a friend and member of a generations-old family of artists, included her in his 1675 history of German art, *Deutsche Academie*:

All kinds of decorations composed of flowers, fruit and birds, in particular also the excrement of worms, flies, gnats, spiders and all such kinds of creature, with all their possible permutations; she showed how each species is conceived and subsequently matures into a living creature, as well as indicating the plants on which they feed. . . . Works like these seemed to emerge from her hands daily.

Sandrart marveled that she was also an excellent housekeeper. She was teaching, too, instructing a *Jungfern Combanny* ("company of maidens") how to paint flowers and successfully experimenting with ways to make colors stay fast on fabrics when they were washed.

Maria Merian Gräffin was still learning and experimenting as an artist, seeking a style of her own, when she made her first foray into publication with a book of flower paintings. She sold the loose sheets, each displaying a flower or two and their names, and even promoted them as patterns for embroidery. They were, however, more than pretty pictures. She was trying out techniques and ways of representing the blooms in ornamental, decorative styles. She showed some of the flowers planted in the ground, others in vases. Some were more stylized, some like pictures in a tulip catalogue. She was making no attempt at this point to show connections between insect and plant, or between caterpillar and butterfly. The first book, published in 1675, was a success. Maria produced another volume in 1677 and a third in 1680, grouping them under the title *Neues Blumenbuch.*

Title page from volume 3 of Maria's Neues Blumenbuch (New Book of Flowers), *1680.*

As Sandrart's mention of "the excrement of worms, flies, gnats, [and] spiders" highlighted, Merian's interest was not confined to the sort of beauty that women who purchased her embroidery patterns would appreciate. For instance, she later described a stroll with a friend. They had expected to admire beautiful insects and flowers but "since we encountered nothing there, we moved to a common weed and found these caterpillars on the white dead-nettle." She took some home and watched at least one change into an exquisite moth.

When Maria and Graff's second daughter, Dorothea, was born in 1678, Maria was putting together a book that she intended to be "a new invention." It has a very long title and description: *The Caterpillar's Wondrous Metamorphosis and Particular Nourishment from Flowers in which for the benefit of explorers of nature, art painters and lovers of gardens though a completely new invention the origin, food and development of caterpillars, worms, summer-birds, moths, flies and other such creatures, including their times and characteristics are diligently studied, briefly described from nature, painted, engraved in copper and published by Maria Sibylla Graff herself, daughter of Matthaus Merian the Elder.* The book, first published in 1679 (there would eventually be three volumes), is often referred to as her *Raupen (Caterpillars)*, the first word of its German title.

Raupen described her search for insects in Nuremberg and those she brought home to watch develop. When she discovered pupae in holes in her house walls, she watched the moths emerge and fed them sugar water. She encountered anomalies: two differently colored caterpillars spun cocoons and came out as identical moths; one pupa produced flies rather than the expected moth; some caterpillars thrived on a plant that was considered toxic. An image in her *Studienbuch* showed growths on a poplar branch with long, thin galls spreading in the veins of the leaves. She drew the aphids that lived inside the galls and the larvae of a fly that preyed on the aphids . . . not a pretty picture. Yet in 1679, when Maria wrote the introduction to her *Raupen*, her faith in God and his care of creation was still radiant:

> *These wondrous transformations have happened so many times that one is full of praise for God's mysterious power and his wonderful attention to such insignificant little creatures and unworthy flying things. . . . Thus I am moved to present God's miracles such as these to the world in a little book.*

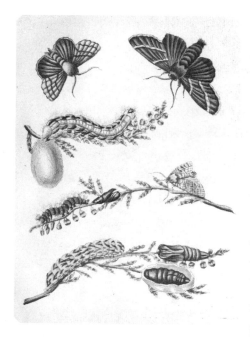

Plate 32 from a 1713 edition of volume 3 of Der Raupen *(Caterpillars), originally published by Maria in 1679. It depicts the life cycle of a moth on a species of the* Erica *(heath) genus.*

Raupen made good on the promise of its title. Maria's interest had shifted almost completely from flowers to insects. Even the roses depicted in *Raupen*'s second volume, published in 1683, would be chewed and infested by insects. The whole drama of insect lifetimes and lifestyles was played out in the first volume's fifty plates, from the laying of eggs on leaves the caterpillar will later eat, to caterpillar, to pupa (again on leaves the butterfly will later eat), to butterfly. It was a stunning, original book, placing each life in the context of other plant and insect life. Nothing like it had been done before, nor would many, except Maria's own, be like it in the future. But she still seemed reluctant to place herself in the same league as her contemporary male scientists. In the introduction to the first volume of *Raupen* she identified herself as a wife working with her husband's permission. Where she might have suggested an explanation for the fact that a fly and moth appeared from the same cocoon, she chose to "leave this to the gentleman scholars."

Maria's thinking was changing in another way as well. Her widowed stepbrother, Caspar Merian, had joined a Pietist religious group led by a man named Jean de Labadie. Pietists preached a simple, devout Lutheranism focused on individual moral choices, rather than on theological debates or church attendance combined with anything-but-Godly lifestyles. Believers didn't require a priest as an intermediary between them and God or Christ or to interpret the Bible for them. Jean de Labadie carried Pietism to extremes, requiring his followers to divest themselves of all worldly goods and worldly pursuits. A group of "Labadists" lived in a closed community on the grounds of a mansion house called Waltha Castle in the village of Wiewert in West Friesland. The mansion had been loaned to three sisters, Lucia, Maria, and Anna Sommelsdijk by their wealthy brother, Cornelis van Aerssen van Sommelsdijk. He was governor of Surinam, a sugar-producing Dutch colony on the northeast coast of South America, where the Labadists had missions and a plantation.

In 1686, for undisclosed reasons, Maria abandoned her marriage of twenty years, left Graff, and joined the Labadist community at Wiewert in Friesland, taking her elderly mother and her two daughters with her. Dorothea was seven, Johanna seventeen. Maria was thirty-nine.

Men and women who joined the community were encouraged, though probably not strictly required, to abandon lives that had been highly intellectual, active, and fulfilling, rid themselves of all their books, and regard all former learning and accomplishment as vanity. However, hardly any abandoned an intellectual life altogether. Instead, they oriented their passions and intellects toward God. All nationalities were welcome, and sermons were translated simultaneously by those who spoke two or three languages. All levels of society were treated as one (though the Sommelsdijk sisters still lived in the mansion, divested of all ostentatious furnishings) with division only being made based on one's success in adhering to the spiritual aspirations of the community. The Labadists lived in poverty, in rooms that were ice-cold in winter, with food frugally doled out, wearing the simplest dress and hairstyles. They toiled in their common fields and gardens to provide their own food, raised livestock, and produced wool to sell. They pooled all their money and possessions, owning nothing as individuals. Taking undue pleasure in one's work was considered vanity, and, to avoid that, assignments of domestic and farming tasks were often made on the basis of what one *couldn't* do well.

Yet Labadie himself was not a joyless man, nor were many of those who chose to follow him. In words that would have resonated with Maria, he expressed his own interpretation of nature:

Everything we hear or see announces God or figures him. The song of a bird, the bleating of a lamb, the voice of a man. The sight of heaven and its stars, the air and its birds, the sea and its fish, the land and its plants and animals . . . everything tells of God, everything represents him, but few ears and eyes try to hear or see him.

Maria was allowed to keep her books and the materials and supplies she needed to continue her work.

In the summer of 1686, Graff arrived at the gates of the compound to fetch his wife. She refused to return to her marriage. Graff wouldn't leave. He lived just outside the gate, did some construction jobs, and sketched the only picture of the compound that has survived. Eventually he declared himself willing to join a religion that seemed alien to him in order to be with his wife and daughters, but the community refused him. When he became ill, Maria visited him but didn't change her mind. Eventually, Graff departed.

Maria wrote that living in Friesland gave her the opportunity to study "what is found specifically in heath and moorland." Her *Studienbuch* tracked her work meticulously. Two years after the summer when Graff had come, her interest turned again to caterpillars, and this time she became more concerned with questions she had not been able to answer before. Some answers were disturbing and, for a woman absorbed in a religious community, were bound to have caused doubts of a loving God. In paintings

she probably produced at this time, some pupae release flies rather than moths or butterflies. Parasitic flies inside the caterpillar or pupa feed on their living host from the inside out until it dies, then hatch in its place. One of Maria's paintings showed a green-and-yellow caterpillar arching up in its death throes as worms emerge from its back, tiny flies swarming around. There would be no butterfly here. In the same picture, she showed the dead, shriveled-up caterpillar. She described such things as "unnatural."

Labadists had founded communities in the New World, one in the colony of Surinam. During Maria's first years at Wiewert, some of these settlers returned ill and dispirited, and even those who stayed in Surinam sent reports of disease, hunger, and mosquitoes. But others also sent moths, lizards, beetles, and fruit, and once even a twenty-three-foot (seven-meter) stuffed anaconda. The species Maria was studying in Europe did not even begin to hint at the variety and marvels of nature that lay waiting in other parts of the world.

At the same time, the community at Wiewert was beginning to fall apart. The horizons were limited and bleak for young people like Maria's daughters, now thirteen and twenty-three. In the summer of 1691, a serious setback rocked the community when many members became ill and died. In September, Maria and her daughters left Wiewert for Amsterdam. Few cities in Europe could have provided such a dramatic contrast with the secluded community where they had lived for ten years.

Inset detail from a map of Amsterdam by Dutch cartographer Frederick de Wit, c. 1688, showing the port of Amsterdam. Maria and her daughters moved to the city in September 1691.

In 1692, Graff, then back in Nuremberg, filed for divorce and remarried. Maria took back her father's name and thereafter signed her work "Merian." Johanna, her older daughter, married that same year. Johanna's husband, Jacob Hendrik Herold, was also a former Labadist but was now engaged in business and had connections in Surinam.

Maria Merian rapidly found means to support herself and her daughters in Amsterdam, marketing fabrics colored in the process that she had discovered could withstand washings, preparing paints for other artists, and selling insects to collectors. Respect and admiration for her artistic and scientific work was growing.

Amsterdam was at the center of world trade, headquarters of the Dutch East India Company. Merian explored the city's botanic gardens at the invitation of its director, visited scholarly collections that included items from Dutch colonies all over the world, and had access to the "cabinets" of some of the world's leading natural philosophers. "In Holland," she wrote,

> I saw with wonderment the beautiful creatures brought back from the East and West Indies . . . the splendid collection belonging to Dr. Nicolaas Witsen, Burgomaster of Amsterdam and President of the East India Company . . . also that of Jonas Witsen, Secretary of the city . . . the collection of Fredericus Ruysch, Doctor of Anatomy and Professor of Botany . . . and belonging to Levinus Vincent and many others.

These were some of the greatest natural history collections in the world, but the stilted presentations displeased her. These collections couldn't show her what she really wanted to know about the insects, "their origins and subsequent development." "All this stimulated me to undertake a long and costly journey to Surinam (a hot and humid land from where the above named gentlemen had obtained these insects) in order to pursue my investigations further."

Merian was well respected but it nevertheless wasn't even worth trying to get anyone to fund a journey like this for a fifty-two-year-old female scientist. She put two hundred fifty-five of her paintings up for sale. She wrote her will. In 1699, she and her daughter Dorothea sailed for Surinam.

Ships headed for South America left port in Amsterdam and reached the Atlantic Ocean through the English Channel, then headed southwest past Portugal, south along the coast of Africa, over the equator, and finally out across the vast Atlantic. Surinam was situated on the northeastern shore of South America, north of the Amazon River and east of the Orinoco. Near this coast, ships were in constant danger of pirates. Merian's managed to reach port safely after two months at sea. In the late summer of 1699, her ship sailed into the mouth of the Surinam River.

Merian was not ignorant of what she was facing. Reports she'd heard of life in Surinam were not inviting unless one were looking for exotic and sometimes dangerous species of animals, birds, and insects, or hoping for wealth from raising sugarcane on plantations that cruelly exploited slaves. Merian expected to spend five years there.

A map of Surinam entitled A New Draught of Surranam upon the Coast of Guianna, *published in London by English cartographer John Thornton, c. 1675, approximately a quarter-century before Maria and her daughter Dorothea set sail across the Atlantic to the exotic South American country.*

Land near the coast was marshy. A few miles inland on rockier ground near the river, European settlers had built the town of Paramaribo. Most of the houses, which sat on a grid of orderly streets paved with crushed shells, were white-frame buildings of two or three stories that looked as though they could have been transplanted directly from the Netherlands. The house of the governor, located just beyond, on the river, had woods and gardens. Past that, a walled fort stood on the river bank: Fort Zeelandia. Farther upriver were sugarcane plantations with elegant homes. Workers on the plantations were mostly slaves brought from Africa, with a smaller number of Amerindians (American-Indian). Merian referred to her African servants as "slaves," though it seems she hired them rather than purchasing them. She also hired an Amerindian woman whom she called "my Indian" and who she eventually brought back to the Netherlands.

Surinam had about a thousand Europeans and almost twelve thousand African slaves at the time of Merian's visit. In addition, and uncounted, were the Amerindian communities in the deep jungles, as well as enclaves of former slaves, known as *maroons*, who also secreted themselves in the rain forest and on tiny tributaries far upriver, having been lucky enough to avoid the brutal, mutilating punishments inflicted on runaways to make future escape attempts impossible.

Merian didn't intend to live at the old Labadist plantation—that was far upriver and nearly defunct. She found a suitable house in Paramaribo, furnished it with supplies, and began seeking permission from plantation owners to visit their lands and look for insects. Life in Surinam took getting used to. The heat was debilitating. Unscreened windows had been no prob- lem in the Netherlands, but in this mosquito-ridden town, they were deadly. Glass or closed shutters made a house far too hot, so windows were instead covered with gauze. Cisterns preserved rain water for drinking and irrigat- ing the gardens but bred more mosquitos. The Amerindians wisely slept in hammocks, which were cooler and avoided contact with vermin and ants that lived on the floor, but the Europeans preferred their sweaty beds. In a letter written after she returned to Amsterdam in 1702, Merian described it:

It is very hot in that country, so that one can only work with the greatest difficulty . . . which is why I could not remain there any longer; also all the people there were amazed that I came out of it alive, for most people there die of the heat.

For Merian, the species she found there were, at least for a while, worth the difficulties. There were more species of insects in her new surroundings than even she could have expected. The overwhelming variety of life and the impossibility of capturing even a small fraction of it were both moving and daunting. She began methodically, making notes and drawings on the small vellum pages of her *Studienbuch* that would later allow her to describe her work and life in Surinam in the introduction to her book *Metamorphosis insectorum Surinamensium* and in the text that accompanied each of its sixty pictures. A typical passage reads:

One day I wandered far out into the wilderness. Here . . . on a tree the residents call the Mispel-Boom [medlar tree] [which] grows to a great height, I found this yellow caterpillar which had pink stripes over its whole body; its head was brown and each segment bore four black spines; its feet were also pink. I took this caterpillar home with me and it rapidly changed into a pale wood-coloured chrysalis, like the one here lying on the twig [shown in her drawing] . . . two weeks later, towards the end of January 1700, the most beautiful butterfly emerged, looking like polished silver overlaid with the loveliest ultramarine, green and purple, and indescribably beautiful; its beauty cannot possibly be rendered with the paint brush.

The first metamorphosis Merian recorded in Surinam, in the fall of 1699, was of a white caterpillar with red knobs on its back and fronds of hair sprouting from the knobs. She found it crawling along a branch of a guava tree, leaving the leaves chewed and ragged. It was not a creature to be touched with bare hands, and Merian on first encounter may have come away with angry pain and swelling, or was cautious enough to wear gloves. Her study notes record that "these caterpillars wrap themselves tightly to the trees." She brought some home and observed "beautiful flies" emerging from some cocoons instead of moths. She would frequently observe the insects that lived in guava trees with foliage being devoured by leaf-cutter ants and huntsman spiders, or harboring the larvae and adults of tobacco hawk moths; or—one of her least pleasant pictures—with a pink-toed tarantula that has killed a hummingbird and is about to attack its eggs.

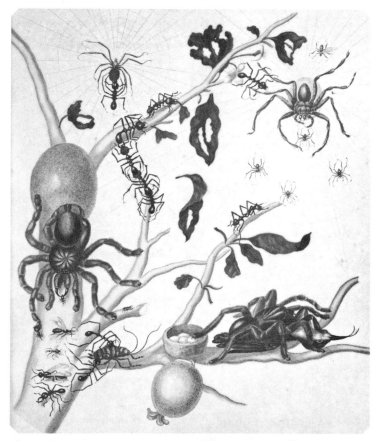

Plate 18 from Metamorphosis insectorum Surinamensium *illustrates the branch of a guava tree (*Psidium guineense *(Swartz)), leaf-cutter ants, army ants, pink-toed tarantulas, huntsman spiders, and a ruby-topaz hummingbird. The pink-toed tarantula has killed the hummingbird and is about to attack its eggs.*

Merian continued to take caterpillars, pupae, and insects found in the wild back to her house and garden, hoping they would survive through all the stages of their lives. She fed them on the kind of leaves on which she had discovered them, watched the caterpillars as they ate and grew and formed pupae, observed a pupa as the adult insect emerged and the adult as it laid eggs for the next generation. She recorded it all in drawings and words.

Merian's Amerindian and African servants were sympathetic to her work, and they carefully brought her living specimens. She learned the local names for plants and insects unknown in Europe. Her servants showed her plants that they used for medicines and told her how certain seeds and fruits could be used to produce body paints, but she recorded nothing about their magical and ritual uses. Perhaps she found them too alien to her Christian beliefs or felt that mention of them would lessen her scholarly credibility in Europe.

One insect that was already well known in Europe was the lantern fly, with a head and long, snout-like projection hugely out of proportion to its body, looking something like an alligator. Lantern flies were coveted by European collectors. Merian's Amerindian acquaintances brought her a good number of them that she put in a box in order to study and watch them develop. Part of a lantern fly's head can at times appear luminous but not fiery bright. Perhaps she kept cicadas (thought at the time to be one stage of lantern fly development) and fireflies in the same box, because she recorded that one night a loud noise woke her entire household—a sound coming from the lantern fly box. When she opened the box, "a fiery flame came out." It's remarkable that the Amerindians brought her the flies at all, for some thought their bite was deadly. Merian drew them with cicadas on a double-blossomed pomegranate.

She painted a peacock flower with a tobacco hawk moth and its larva. She learned that Amerindian women used this flower's seeds to hasten labor in childbirth and, earlier in a pregnancy, to abort babies so that they would not be born into slavery like their parents. Once imported to Europe, the peacock flower would become a garden favorite, its sadder use forgotten, and scholars would scoff at the idea that it could be used in this grim way.

In 1700, Merian ventured beyond the town and the plantations. What remained of the Labadist colony, La Providence, was several days' journey by boat upriver from Paramaribo. In this alien place, the Labadists had apparently lost their minds, and certainly their faith and morality, for they had gained a reputation for being some of the cruelest slave owners in the New World, on the argument that decent treatment of slaves already accustomed to beatings and harsh punishment was achieving nothing whatsoever in the way of obedience or discipline. Merian found only a pitiful remnant of the colony. La Providence did provide, however, a magnificent find: a beige, brown, and violet moth with a wingspan of almost a foot (twelve inches)— greater than any other in the Americas. It was a *Thysania agrippina*.

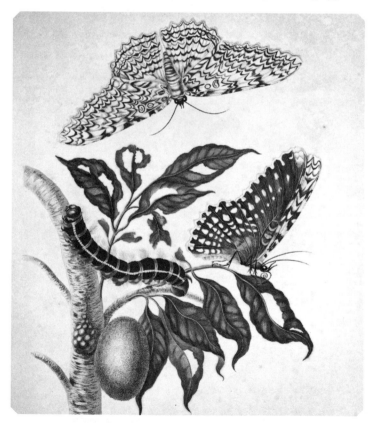

Plate 20 of Metamorphosis *depicts the impressive* Thysania Agrippina, *or birdwing moth—with a wingspan of almost a foot (twelve inches)—on the branches of a gumbo-limbo tree* (Bursera simaruba).

Inevitably, Merian ventured into the rain forest itself. It covered most of the country and was vast and dense enough to conceal the villages of runaway slaves and homes of Amerindians no longer able to inhabit the coasts as they had for centuries before the Europeans arrived. Penetrating the jungle at all was a formidable task and, for a European, a frightening entry into an unwelcoming, alien world. The still air was pierced with eerie wild-bird calls and chilling animal cries. The ground underfoot was watery, crawling with insects and snakes. Enormous ropes of vines, densely tangled, swung from trees growing so close together that it was impossible to pass between. Before entering, Merian sent her servants ahead to hack out a pathway. The animals and insects she longed to observe and record lived in the canopy two hundred feet (sixty meters) above her head, far above the reach and sight of anyone on the ground. The world up there existed for itself alone, mocking human curiosity.

But trees not so tall could be dealt with. Merian sent her helpers up ladders and ordered some trees cut down. When one plant had leaves too delicate to be removed and brought back to her house, her servants dug it up, roots and all, and replanted it in her garden.

Merian didn't confine her interest entirely to insects. One of her pictures showed a dwarf caiman (closely related to the alligator) locked in battle with a false coral snake that is after its eggs. She drew the caiman's scales and the segments of its tail in minute detail, though she couldn't have seen that during their fierce combat. The painting must have been done either from dead bodies or animals that she had in captivity.

Eventually, illness forced Merian to cut short her stay in Surinam—"I almost had to pay for it with my life," she wrote to Johann Georg Volkammer, a medical friend who had published a book about the city gardens of Nuremberg—but the success of having gone there was enormous from both a scientific and business point of view. She explained to Volkammer:

Merian's hand-colored etching, appearing in Johannes Oosterwyk, ed., Metamorphosis insectorum Surinamensium, *1719. Oosterwyk identified it as a dwarf caiman battling a false coral snake. Not every picture that Oosterwyk included in his edition of Merian's book was correctly attributed to Merian, but this one is definitely hers.*

> *While in [Surinam] I painted and described the larvae and cater-*
> *pillars as well as their food and habits; everything I did not have*
> *to paint [immediately] I carried back with me: butterflies, bee-*
> *tles and everything I could preserve in brandy or press I am now*
> *painting the way I used to do when I was in Germany, but on*
> *vellum in a large format, the plants and creatures life size.*

She brought back a crocodile, snakes, iguanas, twenty jars of butter-flies, insects, and fireflies. She was able to sell the crocodile, turtles, two large snakes and eighteen small ones, and other insects. Her illustrations would sell for as much as forty-five florins apiece—the equivalent of about $500 today.

Recovering in Amsterdam and having recouped some of the cost of her passage, Merian took out her *Studienbuch* with the notes, paintings, and sketches she had made in Surinam and set to work on what would be her great book: *Metamorphosis insectorum Surinamensium*, consisting of sixty full-page illustrations and her commentary about each. She would show in meticulously observed detail and vibrant color the life cycles, reproduction, development, and plant environments of caterpillars, moths, butterflies, worms, maggots, beetles, bees, and flies—all in beautifully executed, dynamic paintings, each a separate, exquisite work of art. Merian's style had changed. In these pictures, leaves, stems, and flowers twist and spiral. Plants and insects aren't centered between neat white margins. The riotous abundance of life escapes the boundaries and refuses to obey rules of obvious symmetry. It sometimes seems as though Merian must have craned her neck at uncomfortable angles to see her subjects. Many of these creatures and plants were previously unknown in Europe, and she used the names the Amerindians had given them.

Natural histories written by Merian's contemporaries often included suggestions for medicinal and culinary uses, and Merian followed suit. "One eats [the pineapple] raw or cooked or can make wine or brandy from it." And of the cassava root, "Should a man or an animal drink the extracted juice cold, he or it dies an extremely painful death; but if this water is boiled it makes a very good drink." The root could also be pressed and baked and "has the same taste as a Dutch rusk [Zwieback]."

Merian finished *Metamorphosis insectorum Surinamensium* in April 1705, dedicating the book to "lovers and investigators of nature." In spite of her experience with the violence of the jungle and the "unnatural" parasites that invaded the bodies of living caterpillars, she said in the introduction that she did her work "for the glory of God alone, who created such wonders."

Merian didn't expect anyone to rival her achievement any time soon and wrote in a letter to Volkammer, "this work is rare and will remain rare . . . since the trip is costly and the heat makes living extremely difficult."

The book would place her securely among the scholars of her day, but in her usual deferential manner, its introduction states that

> *because the world today is very sensitive and the learned differ in their opinions, I have kept simply to my observations; in so doing [providing] material for each individual to draw his own conclusions according to his own understanding and opinion, which he can then evaluate according to his own judgement.*

Metamorphosis insectorum Surinamensium was not intended to be a money-making project. Merian knew she would be fortunate merely to cover its costs, for she had decided to make it an extraordinarily beautiful book, a collector's item requiring "the most famous engravers and the best paper so that the connoisseur of art as well as the lover of insects could study it with pleasure and joy." It was a large volume—sixty of the largest-size pages available at the time—and the cost to produce each copy was forty-five florins. To avoid a cash-flow problem, Merian sold subscriptions and advance orders. Always an expert networker and merchandiser of her own work, quick to take advantage of all possible connections, she corresponded with James Petiver, a pro-

digious collector in England, and even suggested that he might introduce her work to Queen Anne of Great Britain. Petiver advertised her books through the *Philosophical Transactions of the Royal Society.* He was no poor merchandiser himself and knew that the oddity of a woman making such an exotic journey was as good a selling point as the paintings. His advertisement read:

Title page of the first edition of Metamorphosis insectorum Surinamensium, 1705.

That Curious Person Madam Maria Sybilla Merian . . . *being lately returned from* Surinam in the West Indies, *doth now pro-pose to publish a* Curious History *of all those* Insects, *and their transmutations that she hath there observed, which are many and very rare, with their* Description *and* Figures *in large* Folio *on* Imperial Paper, *containing* 60 Tables, *curiously performed from her own Designs and Paintings. These she proposes at thirty shillings a* Volume, *viz. ten shillings in hand, and ten more at the receipt of one Moity or 30 Tables, and the rest to be delivered on the third payment.*

The book was an enormous success, highly praised in intellectual circles, bought by the finest collectors, on display in natural history libraries. It also appeared in aristocratic drawing rooms as what we would call a coffee-table book.

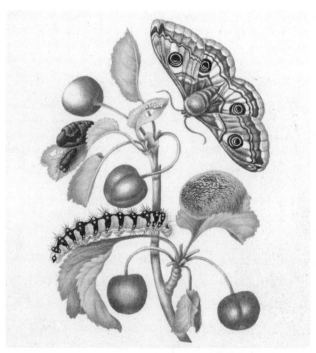

Plate 23 from Der Raupen (Caterpillars), *1679, depicting the life cycle of the small emperor moth* (Saturnia pavonia).

With *Metamorphosis insectorum Surinamensium* finished, Merian decided to produce a third volume of her *Raupen*, the caterpillar book, based on her earlier studies in Wiewert. She was able to find examples of the same species she had encountered near the Labadist community and complete investigations she had begun previously. By this time, she'd definitely decided that the flies she had seen emerging from caterpillar cocoons were not the result of spontaneous generation, and she accepted the cruel parasitic process that produced them.

A young scholar from Frankfurt visiting Merian in 1711 wrote that at age sixty-two, she was "still very lively . . . and hard-working, a very courteous woman." The "hard-working" was not to last. After recording that a berry-eating caterpillar emerged from its pupa, Merian made no more entries in her *Studienbuch*. She stopped work on her third *Raupen*, though in 1714 she oversaw the creation of a Dutch edition of the first two volumes with some additional observations, writing in the introduction about the "government of the Creator, which has put such wonderful life and beauty into such small animals that no painter with a brush and paint could achieve as much." In 1715, she suffered a stroke. She died in early January 1716.

In 1717, Russian Czar Peter the Great visited Amsterdam and hired Dorothea's second husband, Georg Gsell, to help him choose art purchases. The Czar bought most of what remained of Merian's work in the Gsells's possession to become part of his own magnificent art collection in Saint Petersburg. He continued to collect Merian's work whenever it became available. Another set of her paintings from the collection of Sir Hans Sloane eventually went to the British Museum and was hung in Windsor Castle.

In spite of this seeming success, Merian's legacy suffered an unfortunate fate. Various people such as Petiver used her material in ways that would have saddened her. Pictures of individual insects and plants that she had carefully placed together in order to make sense of their lives and transformations were extracted from her works and used in a higgledy-piggledy sort of way, or lined up in strict order. Petiver also translated and revised

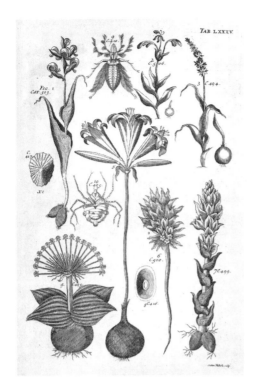

A page from James Petiver's Historiam naturalem spectantia, *1767.*

Merian's written words, giving his own names to categories he created. A new, longer edition of *Metamorphosis insectorum Surinamensium* appeared in 1719, published by Johannes Oosterwyk, who added twelve new plates. It is from this edition that the painting of the caiman and snake in combat comes. Merian had intended it to be part of a reptile book that she never had a chance to produce. Other of Oosterwyk's additions are possibly from another artist and suffer in comparison. Jean Frédéric Bernard in France rearranged the pages of the *Raupen* caterpillar books, added other of Merian's images, took out all her text, and called his book *History of the Insects of Europe*. In these inaccurate reissues, revisions, and many more, the colors in the pictures were sometimes changed or reduced to black and white. As Kim Todd, who traced all these editions, has sadly pointed out, these misrepresentations eventually had a much greater circulation than the originals.

Merian's works undoubtedly continued to influence the representation of nature in art in the eighteenth century, but they were more appreciated as art than as contributions to science. These were the most meticulously drawn and minutely accurate representations of insects, plants, and animals in their environments that anyone produced before the advent of photography, but studying and representing nature in this way was not going to be the wave of the immediate future. It was the more static vision—with drawings of insects lined up in neat rows—that would seem more useful, not paintings so teeming with life that they almost spilled from the paper.

In 1735, when Carl Linnaeus published his book *Systema Naturae,* which catalogued all known life and named all known species, he made ample use of one of the less-reliable, later editions of Merian's books to help identify some species for which he had no specimens. But the study of metamorphosis itself became, for a time, a thing of the past.

Merian's legacy and reputation suffered serious blows in the first half of the nineteenth century. In the 1830s, English naturalist William Sharp Macleay attempted to get a tarantula to eat a hummingbird. His tarantula fled in panic. In 1834, British naturalist Lansdown Guilding published an article called "Observations on the work of Maria Sibilla [*sic*] Merian on the Insects, etc. of Surinam." It was full of snide criticism, condescension, and flat refusal to believe such things as her observation that flies came out of a cocoon rather than a caterpillar. He criticized her for listening to the tales of the Amerindians and believing that the beautiful peacock flower could cause abortions. Another attack came in 1854 from German naturalist Hermann Burmeister, who scoffed that the undeserved popularity of Merian's work had been the result only of its "showy" format. The result was a sharp decline in respect for and trust in Merian's work. She had become "some befuddled lady, dabbling in areas beyond her ken." When Henry Walter Bates, traveling in Brazil with Alfred Russel Wallace, confirmed that tarantulas *do* eat birds (finches, in this case), the news helped restore Merian's reputation, but only a little.

With the publication of Charles Darwin's *On the Origin of Species: By Means of Natural Selection* in 1859, it became clear that studying insects and animals in the environments that shaped them was vital to the understanding of nature. Columns of insect drawings in rigid categories could not tell the whole story. German zoologist Ernst Haeckel invented the term *ecology*. Understanding metamorphosis was essential. Nevertheless, while scientists got busy in the field she had almost single-handedly invented two hundred years before, Merian's work was ignored. Her paintings have appeared in numerous exhibitions, but she is skipped over in books about the history of science where others are credited with discoveries she first made.

As William T. Stearn summed up Merian's genius in the introduction to his translation of *Der Raupen*, few people have ever *seen* nature as Merian did, with the eyes and sensitivity of an exceptionally talented artist, and the careful, scrutinizing eyes of a scientist.

The artist is remembered . . . celebrated. The scientist is all but forgotten.

SIX

Alfred Russel Wallace

THE OTHER DARWIN

(1823–1913)

N THE ISLAND OF GILOLO (modern-day Halmahera) in the Malay Archipelago, the tropical heat was sweltering, but Alfred Russel Wallace lay shivering under a blanket. Wallace was a remarkably resilient man, able to rebound quickly from any disease or disaster the tropics or the rest of the world threw at him, and he lived to age ninety, but in the years when he traveled as a naturalist in the most remote and little-explored parts of the world he grew accustomed to infections and attacks of malaria. Sweat-soaked, trembling, weathering this particular attack in February 1858, he nevertheless used the time well, mulling over a stubborn question: In the ten years he'd spent collecting plants, insects, and animals in South America and the Malay Archipelago, noting the enormous diversity among them, he had become increasingly curious about what mechanism could possibly account for the emergence of new species.

As Wallace recalled later in his memoirs, *My Life: A Record of Events and Opinions*,

> The problem then was not only how and why do species change, but how and why do they change into new and well-defined species, distinguished from each other in so many ways; why and how do they become so exactly adapted to distinct modes of life; and why do all the intermediate grades die out ... and leave only clearly defined and well-marked species, genera, and higher groups of animals?

Photograph of Alfred Russel Wallace, c. 1895.

In 1858, Wallace had most recently been ruminating about ideas put forward by Thomas Malthus* about the checks such as war, disease, famine, and infertility that prohibit unbridled population growth.

> It then occurred to me that these causes or their equivalents are continually acting in the case of animals also; and as animals usually breed much more quickly than does mankind, the destruction every year from these causes must be enormous in order to keep down the numbers of each species, since evidently they do not increase regularly from year to year, as otherwise the world would long ago have been crowded with those that breed most quickly. Vaguely thinking over the enormous and constant destruction which this implied, it occurred to me to ask the question, why do some die and some live? And the answer was clearly, on the whole the best fitted live.... And considering the amount of individual variation that my experience as a collector had shown me to exist, then it followed that all the changes necessary for the adaptation of the species to the changing conditions would be brought about.... In this way every part of an animal's organization could be modified exactly as required, and in the very process of this modification the unmodified would die out, and thus the definite [sic] characters and the clear isolation [sic] of each new species would be explained.

In Gilolo, there were two hours between the onset of Wallace's chills and their disappearance. He was left in a pool of sweat and with the entire theory of natural selection in his head. Weak and exhausted, he sketched out his answer that evening, and in the following two evenings he wrote down his theory in detail in an essay with the lengthy title "On the Tendency of Varieties to Depart Indefinitely from the Original Type; Instability of Varieties Supposed to Prove the Permanent Distinctness of Species." In early March, when he was happy with his essay, Wallace sent it off to

---

* Malthus's theory of population.

Photograph of Charles Darwin taken c. 1855, for the Literary and Scientific Portrait Club.

Charles Darwin in England and then sailed to New Guinea to continue his exploration and collecting.

Wallace's essay—fewer than a dozen pages describing his new theory—reached Darwin in June with Wallace's accompanying letter suggesting that Darwin might find these ideas helpful in his own work. "Helpful" was almost certainly not the first word that sprang to Darwin's mind. Wallace's essay put forward the same theory that Darwin had been working toward for two decades but hadn't yet published or discussed with anyone except botanist Joseph Hooker and geologist Charles Lyell. As Darwin lamented to Lyell, "I never saw a more striking coincidence. If Wallace had my . . . sketch written out in 1842 he could not have made a better short abstract." Wallace had discovered the secret of the origin of species.

Many of us may not be familiar with this story, but we know how it ends. It is Darwin who is acclaimed as one of the greatest discoverers in the history of science and knowledge. Wallace has been sidelined. It is impossible to escape the conclusion that this was how Wallace wanted it, that he colluded in his own obscurity.

Wallace's background was dramatically different from Darwin's. He didn't come from a well-off intellectual family or have a university education. But both men's obsession with nature began early. Wallace was fourteen when he started working with his older brother William as an apprentice land surveyor. Hiking over immense properties, exploring the countryside, and reading William's stock of books and magazines, Wallace began to appreciate maps and the way borders and boundaries are determined, acquired a knowledge of basic geology, and became enthralled with plant and animal life. He was unfailingly optimistic, impossible to discourage, strongly self-motivated. While his brother went looking for jobs, Wallace studied, learned what he pleased at his own pace, and began to develop into the independent thinker he would be as an adult, with complete lack of concern or even awareness about whether he was unconventional or unorthodox. Never a dabbler, he was satisfied, even as a child, only with complete mastery of any subject he took on.

In threadbare clothes because money was so short, the two brothers moved from town to town for surveying jobs. The itinerant life was no problem for Wallace. He would never live in any one place for more than a few years for the rest of his life. After a move to Neath, in Wales, Wallace became not only an appreciator of nature but a collector. He learned to preserve wildflower specimens and, eager to understand the patterns he was starting to realize underlie nature, he began a new self-education. He sought out books on botany and taxonomy: the sciences of classifying plants and animals. When he was nineteen he purchased William Swainson's *A Treatise on the Geography and Classification of Animals*. For Wallace, reading Swainson's treatise inspired what would become a lifelong

curiosity about how and why animals are geographically distributed as they are.

What possible future was there in all this? Traveling as a naturalist was a way to spend your fortune if you were independently wealthy, but Wallace was far from that. In 1843, he took a position teaching drawing, surveying, and mapping at a church school in Leicester.

Undated photograph of the English naturalist Henry Walter Bates.

There in the Leicester library he found Thomas Malthus's *An Essay on the Principle of Population* and encountered the ideas that much later would preoccupy him on his sickbed and trigger his discovery of natural selection. Also in Leicester, he met a young man about his age named Henry Walter Bates. Bates was a self-motivated scholar like Wallace, with only slightly more formal education. In fact, he had already published a paper on beetles. Bates introduced Wallace to the world of insects, and Wallace began to realize the seemingly infinite variability of species; the almost endless variety of structures, shapes, colors, and markings that made one distinguishable from another; and the way they had adapted to different conditions.

Wallace and Bates shared a dream: to travel to the Amazon rain forest in South America in search of botanical and zoological specimens. It was a ridiculous plan, the romantic fantasy of two young men with dangerously inflated opinions of their own capabilities and invulnerability. They had no formal education in their area of interest and no background in taxidermy. They were

A plate from The Narrative of the Surveying Voyages of His Majesty's Ships *Adventure and* Beagle . . . , *published in 1839, showing a surveying party approaching Button Island in the Tierra del Fuego archipelago.*

almost completely ignorant about how to capture and skin birds and mammals. They had no skills to survive in a hostile jungle or in an unreliable boat on a river swarming with alligators. Neither of them had credentials or official sponsorship, rich patrons, independent wealth, or families able to underwrite their journeys (as Darwin's father did for him). They couldn't pay their passage by serving as medical officers and naturalists on board a ship. They might be able to cover some of their expenses by selling extra specimens through agents back in London, but the main business of the adventure would not be to make money in that way; it would be to gather facts that would help explain the origin of species—the holy grail for philosophical naturalists.

Darwin's *The Voyage of the Beagle*, Alexander von Humboldt's *Personal Narrative of a Journey to the Equinoctial Regions of the New Continent*, and William Henry Edwards's *A Voyage up the River Amazon: Including a Residence at Pará* had given Wallace an unquenchable thirst to go where few Europeans had ventured and make discoveries that would be significant contributions to scientific knowledge.

Throughout his life Wallace was absolutely impossible to discourage, and Henry Bates was his match. They studied collections in the British Museum and the Royal Botanic Gardens, Kew. Neither institution was willing to support their expedition, but both were interested in anything novel that they could send back. They met William Edwards, whose book had inspired Wallace, and who advised them to get to Pará in Brazil by the beginning of the dry season in late May. They consulted a curator at the India Museum in London for advice on how to store and transport natural history specimens. They contacted a "Mr. Stevens," who was known to be a helpful, honest agent for many traveling naturalists. Stevens advertised and auctioned specimens to scientific societies and collectors and kept his wandering clients informed about what was in demand back in Britain.

To finance their trip and make any sort of profit, Wallace and Bates would have to send Stevens thousands of specimens. Careful preservation of specimens would be difficult and time-consuming but also key to their success. A badly preserved specimen was worthless and represented lost revenue and effort. Stevens advised them to keep field notebooks with extensive information about each specimen: the location it was found, the climatic conditions, and as much as they could discover about its habits, habitat, and environment. Stevens's shop near the British museum sold supplies: knives, scissors, scalpels, pliers, pins, needles, hammers, hatchets, cotton, paper, nets of various sizes and types, forceps, phials, pillboxes, etc. Novices in their use, Wallace and Bates would learn the necessary techniques by trial and error in the field. Papers and passports were another hurdle; but finally, on April 26, 1848, they boarded the *Mischief* in Liverpool to sail to Brazil. Wallace was twenty-five years old.

Thirty years later, recalling his first venture into the true, virgin equatorial forest, Wallace described

> . . . *weird gloom and solemn silence, which combine to produce a sense of the vast—the primeval—almost of the infinite. . . . Man seems an intruder . . . overwhelmed by contemplation of the ever-acting forces which, from the simple elements of the atmosphere, build up the great mass of vegetation which overshadows and almost seems to oppress the earth.*

Wallace had one advantage over other novice naturalists: his impoverished childhood and itinerant existence. Living in poor—even primitive—accommodations as a surveyor's apprentice and never growing accustomed to a soft life made him better prepared than most for the life he was so determined to encounter.

An illustration from The Naturalist on the River Amazons . . . *by Henry Walter Bates, 1863.*

151

An engraving of the umbrella bird from Bates's The Naturalist on the River Amazons.

From the summer of 1848 until July 1852, Wallace—first with Bates and later alone or with his brother Edward—trekked far into the remote, unexplored backcountry of Brazil and the Amazon watershed, all the way to the tributaries in Venezuela of the Rio Negro, more than halfway across the continent. The farther Wallace went, the more daring he became, traipsing at night through areas rife with poisonous snakes and jaguars, teetering on narrow log bridges to cross torrents and sometimes slipping off, suffering near-deadly bouts of malaria. He was able to collect many previously unknown species and varieties, as well as others, already known, that collectors prized back in England. Again and again he encountered the puzzle of why the geographic distribution of some species was so precisely restricted and unexplainable. Wallace's article about the umbrella bird, published in *Proceedings of the Zoological Society of London* in 1850, was his first publication in a scientific journal.

It was a hard-won and astounding collection that Wallace loaded aboard the ship *Helen* on July 12, 1852, for the return to England. Records show ten thousand birds skinned and ready to be mounted, thousands of Brazilian plants, more birds' eggs than any museum could boast, and even live animals.

Just short of a month at sea, Wallace tragically lost it all. The ship, which was carrying highly flammable palm oil, caught fire. Everything had to be abandoned—his notebooks, his specimens, his live animals . . . all except his tin box, which contained a few shirts, a watch, a little money, his drawings of fish and palms, and his journal and notes from his travels on the Rio

Negro. Frantically bailing in leaky lifeboats surrounded on a rough sea by burning fragments, crew and passengers watched the ship go up "in a most magnificent conflagration." A few of Wallace's live animals hung on to the finish, clustered on the *Helen's* bowsprit, but they couldn't be coaxed into a lifeboat and were incinerated except for one fortunate lone parrot that fell into the sea, managed to cling to a piece of debris, and was saved. Wallace was so involved in the struggle for survival that it was some time before he was able to recognize what an enormous personal disaster he had suffered.

> *All my private collection of insects and birds . . . was with me, and comprised hundreds of new and beautiful species, which would have rendered (I had fondly hoped) my cabinet, as far as regards American species, one of the finest in Europe. But besides this I have lost a number of sketches, drawings, notes, and observations on natural history, besides the three most interesting years of my journal, the whole of which, unlike any pecuniary loss, can never be replaced. How many times, when almost overcome by the ague, had I crawled into the forest and been rewarded by some unknown and beautiful species! How many places, which no European foot but my own had [ever touched]! How many weary days and weeks had I passed, upheld only by the fond hope of bringing home many new and beautiful forms from those wild regions. . . .*

The journey to South America had started as an impossible dream. It was, once again, only a dream.

The *Helen* sank in an area where ships passed frequently, but the supply of food and drink in the lifeboats was almost depleted when her crew and passengers were picked up days later by a ship that had scant rations itself. Having survived a hurricane and a waterspout, the badly damaged *Jordeson* finally toiled into the English Channel only to encounter another violent storm. Other ships were lost, but the *Jordeson* landed at Deal on October 1, 1852. Wallace had been at sea for eighty days.

Three days later, Wallace tottered into the Royal Entomological Society of London, a pitiable figure with joints swollen from malnutrition. He told of his adventures, what he had discovered, and what he had lost. His agent, Stevens, may have encouraged him to make this appearance, which left a profound impression. Here was a survivor whose stout English heart and deep devotion to his science embodied Victorian manhood and science at its best. Stevens gave Wallace a check for £200, the amount for which he had insured Wallace's collection.

Wallace did his best with what had survived, publishing papers based on his few notes, his memory, and letters he had written from Brazil, and attending, as a non-member, meetings of the Royal Entomological Society, the Zoological Society of London, and the Royal Geographical Society. He wrote two books, his interests again turning to the unexplainably limited geographical ranges of some species and the way species change over time.

As his next destination, Wallace chose the Malayan Archipelago. Except for those islands occupied by the Dutch, the vast Archipelago was mostly unexplored, and there were tales of exquisite beauty and mineral wealth. He sailed on March 4, 1854, and at the end of May he shipped seven hundred species of beetles to Stevens from Singapore. On this expedition, Wallace kept scientific societies and journals up to date on his activities.

In February 1855, after a rainy season during which he'd had time to think more about the questions that were "rarely absent from my thoughts," Wallace penned an essay entitled "On the Law which has Regulated the Introduction of New Species." The paper introduced his "Sarawak Law": "Every species has come into existence coincident both in space and time with a pre-existing closely allied species." According to this law, there have been no breaks or sudden appearances, either in time or in space. Wallace's Sarawak Law wasn't the answer to all his questions, but it was a stop on the way there. As he wrote,

The evidence pointed to some kind of evolution. It suggested the when and where of its occurrence, and that it could only be through natural generation; but the how was still a secret.

A page from Wallace's notebook on birds in Macassar, c. 1856; the hornbill bird (Buceros cassidix) is shown here.

Wallace thought it was fairly obvious that organic life found in the present world was the result of a natural process of gradual extinction and creation of species from the most recent geological periods. It seemed logical to infer that this natural sequence had also been happening much earlier, in one geological epoch after another.

The fact that closely related species tend to share habitats or live in adjoining habitats was an important clue that Wallace could verify with examples from his own experience. Also, in geological formations, there was evidence of extinct species related to currently existing species. Considered either in space or time, there was continuity. This continuity hinted at descent from some common ancestor, with gradual appearances of new forms that branched from that ancestor.

Wallace suggested thinking of a tree—a vertical trunk with diagonal branches—to show how orders, families, genera, and species had developed, with many coming from a common ancestral form; and how each of these species may in turn have been the ancestral form for other closely related species. Wallace would also point out how far we are from being able to see this complete branching "tree."

> *If we consider that we have only fragments of this vast system, the stem and main branches being represented by extinct species of which we have no knowledge, while a vast mass of limbs and boughs and minute twigs and scattered leaves is what we have to place in order, and determine the true position originally occupied with regard to the others, the whole difficulty of the true Natural System of classification becomes apparent to us.*

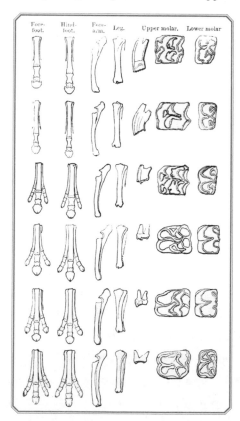

An illustration titled "Geological development of the horse tribe" from Wallace's book Darwinism: An Exposition of the Theory of Natural Selection, *1891, demonstrating evolution. Wallace writes that the "transition from the Eocene* Eohippus *to the modern* Equus *has taken place in the order indicated."*

In his paper on the geographical distribution of organisms on earth, Wallace referred to Darwin's discoveries in the Galápagos Islands: If an area had species, genera, or even whole families not found anywhere else, then that area must have been isolated long enough for many series of species to have been created from preexisting ones which have become extinct. He hypothesized that the explanation for what Darwin had found on the Galápagos Islands was that ancestral types of the present-day species had arrived, as on other newly formed islands, "by the action of winds and currents, and at a period sufficiently remote to have had the original species die out, and the modified prototypes only remain."

The self-assured yet always self-deprecating Wallace coached all this as an invitation.

> *The writer's object in putting forward his views in the present imperfect manner is to submit them to the test of other minds, and to be made aware of all the facts supposed to be inconsistent with them. As his hypothesis is one which claims acceptance solely as explaining and connecting facts which exist in nature, he expects facts alone to be brought to disprove it, not à priori arguments against its probability.*

"On the Law which has Regulated the Introduction of New Species," also known as the Sarawak Essay, was a beautiful paper, redolent with the overwhelming, awe-inspiring richness of the present and past natural world, as well as of Wallace's deep understanding of it and his suspicions of what still lay beyond anyone's knowledge. Yet when the paper appeared in the September 1855 issue of the *Annals and Magazine of Natural History*, there was distressingly little response. Wallace wrote to Darwin in September 1857, expressing disappointment that his paper "had neither excited discussion nor even elicited opposition" and requesting some feedback. He also enquired whether Darwin intended to address the question of the origin of human beings in a forthcoming book.

Darwin was one of the few who *had* read Wallace's essay. The notes he jotted in his copy were generally favorable, though he claimed the simile of the tree was his own. Geologist Charles Lyell also had read Wallace's essay. In April 1856, Lyell visited Darwin, who for the first time told Lyell of his idea that natural selection was the mechanism of evolution. Lyell urged Darwin to publish this immediately. Wallace's essay was ample evidence that Darwin and Wallace were thinking alike, and Lyell feared that Wallace might publish first and claim priority. Darwin insisted he hadn't yet gathered enough facts. Again, a few weeks later, Lyell urged Darwin him to "publish some small fragment of your data." Darwin replied, "I do not know what to think. . . . I certainly should be vexed if anyone were to publish my doctrines before me," yet he didn't want to publish solely for the sake of priority. He decided not to follow Lyell's suggestion but instead to write a book that would organize the data he had collected over eighteen years and make a strong argument. It took Darwin another five months to get started on "Natural Selection."

Meanwhile, Wallace was continuing his travel and collecting, sending nine thousand specimens of sixteen hundred species back from Macassar—this in spite of ants that made colonies in his bird skins—"whence they sally out to gnaw the eyelids, the base of the bill and the feet"—destroying the results of weeks of exploration and work. Spiders, larvae, mites, dogs, damp, roaches, rats—all attacked his specimens. How he also found time to write prolifically is difficult to imagine. Praise at the February 23, 1858 meeting of the Zoological Society from John Gould, Britain's leading ornithologist, for "showing great perseverance and energy" seems an understatement.

Lyell had reason to fear for Darwin. With his Sarawak Law, Wallace had skated very near the full discovery of natural selection as the "mechanism" of the origin of species. It would require only a minute leap of thought to get him there. In the late winter of 1858, Wallace sailed to a wild and largely unexplored island: Gilolo. In late February he suffered through his feverish hours, recovered with the theory of natural selection clear in his head, and penned his essay "On the Tendency of Varieties to Depart

*One of Wallace's preserved specimens from New Guinea: a yellow-bellied longbill (*Toxorhamphus novaeguineae *[Lesson, 1827]).*

Collect: *A. R. Wallace*
Donat:

Ingekomen: *W 65.*   Register n⁰. RMNH 133/15

Indefinitely from the Original Type, Instability of Varieties Supposed to Prove the Permanent Distinctness of Species." In early March, he sent it to Darwin.

Darwin had first conceived of the idea of natural selection twenty years earlier, and in the summer of 1844 he'd written a 230-page sketch that only Joseph Hooker had read. Darwin told his wife Emma to publish the pages if he should die before writing a larger book. In 1857, he sketched his theory and how the principle of divergence applied in a letter to botanist Asa Gray at Harvard University.

Earlier in this chapter, we touched only briefly upon the fateful exchange of letters between Wallace and Darwin in 1857 and 1858. In more detail, here is what happened and how Wallace was sidelined: In late May or early June 1858, Wallace's letter and essay about his discovery of natural selection reached Darwin. Darwin sent Wallace's essay to Lyell with a letter that lamented, "Your words have come true with a vengeance. . . . So all my originality will be smashed." Wallace hadn't asked Darwin to publish the essay, "but," Darwin's letter to Lyell continued, "I shall, of course, at once write and offer to send it to any journal."

A week later Darwin wrote again to Lyell, outlining the evidence he had that the idea of natural selection was originally his: his 1844 sketch and the sketch he had sent to Asa Gray a year ago, in 1857. He insisted there was nothing in Wallace's essay that he, Darwin, had not written out more fully in those sketches. No one, upon seeing these, could accuse him of stealing anything from Wallace—of this he was certain. Nevertheless, Darwin felt that to publish his ideas at this point purely to defeat Wallace would be dishonorable.

> *I would far rather burn my whole book than that [Wallace] or any other man should think that I had behaved in a paltry spirit. Do you not think his having sent me this sketch ties my hands? I do not in least believe that he originated his views from anything which I wrote to him.*

Darwin asked Lyell to share the problem with Hooker, and within the week, Hooker urged Darwin to pull together material to present to the Linnean Society at their meeting just three days off. Darwin could not comply. His infant son Charles had just died of scarlet fever, and his wife and daughter were seriously ill with diphtheria. He suggested that Hooker and Lyell present the Society with Wallace's essay, his own sketch from 1844, and the sketch sent to Asa Gray.

British botanist Joseph Dalton Hooker in an 1851 lithograph.

Hooker and Lyell introduced their presentation with a letter containing the words,

> These gentlemen having, independently and unknown to one another, conceived the same very ingenious theory to account for the appearance and perpetuation of varieties and of specific forms on our planet, may both fairly claim the merit of being original thinkers in this important line of inquiry; but neither of them having published his views, though Mr. Darwin has for many years past been repeatedly urged by us to do so . . . we think it would best promote the interests of science that a selection from them should be laid before the Linnean Society.

The Society's undersecretary read the essay and sketches in the order of their dates of composition. There was hardly any discussion. At the end of the year, the president of the Society, Thomas Bell, commented in his Presidential Address that 1858 was not "marked by any of those striking discoveries which at once revolutionize the department of science on which they bear."

Darwin had already begun composing a letter to Wallace relinquishing all priority, but he changed his mind when he heard how strategically Lyell and Hooker had presented his work—making sure the three items were read in the order in which they had first been written, meaning that both of his own sketches were read first, before Wallace's piece. The balance had been tipped. Darwin decided to lay aside his plan to write a lengthy book. At a more rapid pace than he had written anything in his life, he set to work on a version that was shorter and more like an abstract. ". . . It is really impossible to do justice to the subject, except by giving facts on which each conclusion is grounded," he wrote to Lyell. Nevertheless, he admitted "I am almost glad of Wallace's paper for having led to this." Darwin was sure Wallace was writing a book of his own.

Wallace was doing nothing of the sort, though he had in mind a future book presenting his ideas and defending them more fully and clearly. After posting his bombshell to Darwin, he had set sail on March 10, 1858, to Dorey in New Guinea to continue his exploring and collecting.

When letters from both Darwin and Hooker reached Wallace in Ternate in September, Wallace's reply showed that he felt honored to have received a letter from Hooker, whom he had never met:

> *Allow me in the first place sincerely to thank yourself & Sir Charles Lyell for your kind offices on this occasion, & to assure you of the gratification afforded me both by the course you have pursued, & the favourable opinions of my essay which you have so kindly expressed. I cannot but consider myself a favoured party in this matter, because it has hitherto been too much the practice in cases of this sort to impute **all** merit to the first discoverer of a new fact or a new theory, & little or none to any other party who may, quite independently, have arrived at the same result a few years or a few hours later.*

Wallace seems never to have had any doubt that Darwin's ideas were more complete than his own and had preceded his. He wrote to his mother that Darwin had showed his essay to Hooker and Lyell, "who thought so highly of it that they immediately read it before the Linnean Society. This assures me the acquaintance and assistance of these eminent men on my return home." Three weeks later, still basking in what seemed to him a triumph, he wrote to Stevens:

> *An essay on varieties which I sent to Mr. Darwin has been read to the Linnean Society . . . on account of an extraordinary coincidence with some views of Mr. Darwin, being written but not yet published, and which were also read at the same meeting. If these are published . . . send me three, and . . . send one to Bates, Spruce and any other of my friends who may be interested.*

When the papers were published in August 1858—grouped together and in the order they had been presented—in the *Journal of the Proceedings of the Linnean Society of London*, Wallace proudly advised a friend in London, George Silk, to borrow a copy from a member of the society and look for his article, and "some complimentary remarks therein by Sir C. Lyell and Dr. Hooker, which (as I know neither of them) I must say I am a *little* proud of."

In the late autumn of 1858 and in a state of high exhilaration, Wallace could think of nothing but continuing his research, hoping for more discoveries that would underpin his and Darwin's joint theory. In January Darwin wrote a response to Wallace's letters to himself and Hooker. "Permit me to say how heartily I admire the spirit in which [your letters] are written. Though I had absolutely nothing whatever to do in leading Lyell and Hooker to what they thought a fair course of action, yet I naturally could not but feel anxious to hear what your impression would be."

So it would continue for many years. Wallace never mentioned in public or in his writings that he felt he had been cheated. There would later be serious disagreements between Wallace and Darwin but never about who should have been given priority.

A page from the Journal of Proceedings of the Linnean Society of London, *which introduced the simultaneously published versions of Darwin's and Wallace's groundbreaking papers about their discoveries regarding natural selection, published August 1858.*

An illustration entitled "The 'King' and the 'Twelve Wired' Birds of Paradise" from volume 2 of *Wallace's* The Malay Archipelago: The Land of the Orang-Utan and the Bird of Paradise, *1869.*

Wallace would stay in the Malayan Archipelago, traveling among the islands, exploring, and collecting until the spring of 1862. A return to England was impossible until he had sufficient money to live there, and the sale of specimens was the best way to earn it. Meanwhile, his reputation as a naturalist was growing. There was now no problem with having his papers received and read before the Linnean Society—at least, not when he sent them through Darwin. In a paper read before the Society on November 3, 1859, Wallace had written as though evolution were an established fact. He was always a less restrained man than Darwin when it came to making bold statements.

In early April 1859, Darwin had reported to Wallace that his book was ready for publication and that he would mention Wallace's two important papers—"You will, I hope, think that I have fairly noticed your papers in the *Linnean Transactions*"—making it clear that Wallace's explanation of the law was the same as that he was offering. Darwin mentioned that he had, in an earlier unsent letter, promised not to publish before Wallace. He was reneging on that unspoken promise, having changed his mind on the urging of Hooker and Lyell. By this time Wallace was no longer planning a book of his own on natural selection. In an article published in April 1860 in *Ibis* (a just recently established journal of avian science) titled "The Ornithology of Northern Celebes," Wallace spoke of natural selection as Darwin's theory, referring to "Mr. Darwin's principle of 'natural selection'."

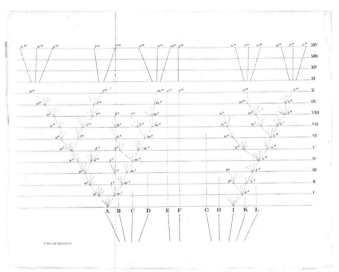

A tree diagram by Darwin illustrating the divergence of species, from his On the Origin of Species, *1859.*

On the Origin of Species was published on November 24, 1859. Darwin sent a copy to Wallace asking for his impression, "as you have thought profoundly on the subject, and in so nearly the same channel with myself." His letter continued, "I do not think your share in the theory will

be overlooked by the real judges, such as Hooker, Lyell, Asa Gray, etc."
It was unlikely Hooker and Lyell could have overlooked it even if they'd
wanted to.

Wallace was awed by Darwin's book. *Origin of Species* would, he wrote
to Silk, "live as long as the *Principia* of Newton." And to Bates,

> *However much patience I had worked up and experimented on
> the subject, I could never have approached the completeness of his
> book—its vast accumulation of evidence, its overwhelming argu-
> ment, & its admirable tone and spirit.*

Darwin was deeply impressed by Wallace's

> *generous manner. . . . Most persons would in your position have
> felt bitter envy and jealousy. How nobly free you seem to be of this
> common failing. But you speak far too modestly of yourself; you
> would, if you had had my leisure, have done the work just as well,
> perhaps better, than I have done it.*

Wallace learned, with the usual time lapse inevitable in his sea-going
correspondence, about the controversy in England over Darwin's book.
Editors of magazines and journals read by the educated public were
almost uniformly hostile. Meanwhile, Wallace continued his work on the
other side of the world, interested in what was going on in England but
largely immune to any attack himself.

When Wallace returned to England in the spring of 1862 after eight
years in the Malay Archipelago, he was a famous and highly respected man
even among those who disagreed with his theory of the origin of species.
Magazine editors and scientific colleagues of the highest order lauded him.
He had collected 125,660 specimens, including 310 mammals, 100 reptiles,
8,050 birds, 7,500 land shells, 13,100 butterflies and moths, 83,200 beetles, and 13,400 other insects—a great many of these previously unknown.
Biographer Ross Slotten has aptly compared this feat with "scaling Mount
Everest or trekking across the South Pole."

A case of Asian beetles collected by Wallace, now in the Natural History Museum of London.

Before long, Wallace was organizing his collections, making contacts with museums and experts, and offering his opinions and presenting papers at the Zoological Society of London, the Royal Entomological Society, the Royal Geographical Society, and the Linnean Society. The Zoological Society accepted him as a fellow. A vivid writer and dramatic speaker, he held audiences in thrall with stories of his adventures and discoveries. The Royal Society awarded Wallace its coveted Royal Medal in 1868, citing mainly his discovery of the "Wallace Line." While exploring the Malay Archipelago, Wallace had decided that the strait between Bali and Lombok (in the Australian part of the Archipelago) was, and probably always had been, a barrier that land animals and most birds could not pass—a theory he came up with on the stark biological contrasts of species he found. The nominating committee called this discovery his most important achievement, and the border is still called the Wallace Line, or Wallace's Line, today.

At age forty-one, four years after his return to England, Wallace married Annie Mitten, who was twenty-three years younger than he, and they began to raise a family. Wallace went on publishing at a prolific pace—he would rack up twenty-one books and about seven hundred essays and articles in his lifetime. He and his wife followed the pattern he had fallen into early in his life: never staying in any one place for long. At first they were in central London, then Barking, then Grays in Essex, then Dorking in Surrey, then Croydon, then Godalming, then in Dorset, in Parkstone, and finally

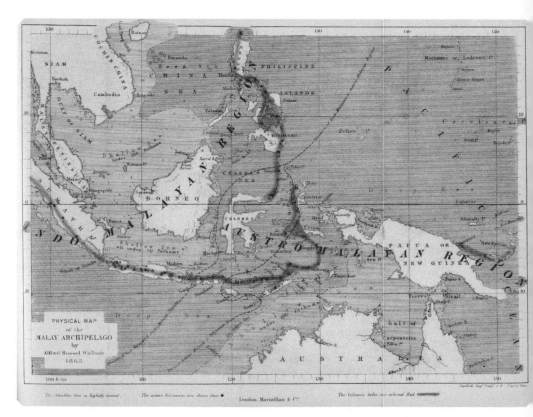

An 1863 map by Wallace titled "Physical Map of the Malay Archipelago" showing the "Wallace Line" from his 1869 book The Malay Archipelago. *Volcanic belts are highlighted in red on the original map (seen in dark gray here).*

Broadstone. Wallace built three of the homes, and he and Annie always had gardens with exotic plants collected for them by friends in various parts of the world.

Sometimes the moves were dictated by economics. Wallace discovered that writing books that were widely bought and read (his *The Malay Archipelago: The Land of the Orangutan, and the Bird of Paradise* in 1869 was particularly successful and is still in print today) could keep him about as well as selling specimens had but couldn't support a family. He supplemented his income by grading school exams, and when at age sixty-three he undertook a year-long, highly successful lecture tour of the United States, that too was largely motivated by the desire to bring in much-needed cash. Wallace and his family often barely skirted bankruptcy and occasionally accepted help from more solvent friends. He sold his collection of exotic birds, one of the finest in the world, to the British Museum.

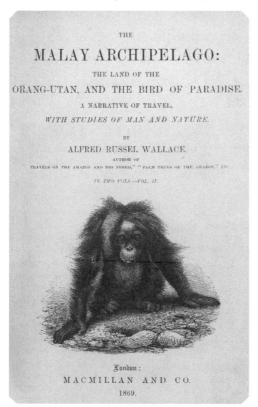

THE

MALAY ARCHIPELAGO:

THE LAND OF THE

ORANG-UTAN, AND THE BIRD OF PARADISE.

A NARRATIVE OF TRAVEL,

WITH STUDIES OF MAN AND NATURE.

BY

ALFRED RUSSEL WALLACE,

AUTHOR OF

"TRAVELS ON THE AMAZON AND RIO NEGRO," "PALM TREES OF THE AMAZON," ETC.

IN TWO VOLS.—VOL. II.

London:

MACMILLAN AND CO.

1869.

In 1881, Darwin, T. H. Huxley (the man known as "Darwin's bulldog"), and others pulled strings to get Wallace on the Civil List, with Darwin devoting a good two months almost exclusively to this undertaking. Wallace learned the day before his fifty-eighth birthday that he would receive a civil pension of two hundred pounds a year—approximately $30,000 today.

The title page of Wallace's The Malay Archipelago.

For most of the 1860s, Wallace and Darwin fought side by side on behalf of their theory in a hostile environment, but by the end of the decade the most powerful people in the English scientific community were agreeing with them. Wallace lamented that there were few opponents left to engage in interesting disputes, and that it might be necessary to propose an outrageous theory just to start a good fight. The only enjoyable remaining controversy was between himself and Darwin.

The causes of the most significant disagreements between them were evident in essays in Wallace's published collection titled *Contributions to the Theory of Natural Selection*. Wallace had reluctantly concluded that natural selection could not have produced the higher faculties of human beings. Human consciousness, the human brain, hands, vocal organs that allow articulate speech, and those faculties that enable humans "to realize the wonderful conceptions of mathematics and philosophy, or give us an intense yearning for abstract truth"—all these were "not explicable on the theory of variation and survival of the fittest." Wallace believed they were present in our most primitive ancestors long before they had any value in the struggle for survival. Darwin disagreed "grievously." Wallace had discarded belief in God and the teachings of the church as a youth and refused to have his children baptized, but he found human consciousness unexplainable by any theory so far. It was a great mystery and seemed to require (he considered this necessity a "disadvantage" of his theory)

> *that a superior intelligence has guided the development of man in a definite direction, and for a special purpose. . . . It is probable, that the true law lies too deep for us to discover it; but there seems to me, to be ample indications that such a law does exist, and is probably connected with the absolute origin of life and organization.*

Other evolutionists were appalled, fearing that arguments within their ranks gave ammunition to those who still opposed Darwinian evolution, but Wallace denied forcefully that he was backsliding on the theory of evolution. He refused to believe that his conclusions in any way weakened the power of the law of natural selection any more than it was weakened by the existence of breeds of domesticated animals and birds whose existence owed to something beyond undirected survival of the fittest.

Another disagreement between Darwin and Wallace was over how important "sexual selection" is in the process of natural selection. Darwin thought features like bright colors were inexplicable except as an advantage in attracting a mate. Wallace favored a broader view: survival mechanisms of many kinds played deciding roles, with favorable variations often so minute, subtle, or hidden that observers could not detect them or understand how they aided survival. He thought protective coloration played a greater role than sexual selection, and he pointed out that even asexual creatures such as caterpillars had brilliant colors.

Though no dispute ever arose between him and Darwin over priority in the discovery of natural selection, Wallace was saddened that his own discovery had been dismissed so easily and that he had been classed with others who'd had similar thoughts but not realized their significance. In the collection *Contributions to the Theory of Natural Selection*, which reprinted his two crucial essays from 1855 and 1858, he wrote,

> *I saw at the time the value and scope of the law which I had dis-*
> *covered, and have since been able to apply it to some purpose in*
> *a few original lines of investigation. But here my claims cease. I*
> *have felt all my life, and I still feel, the most sincere satisfaction*
> *that Darwin had been at work long before me, and that it was not*
> *left for me to attempt to write* The Origin of Species. . . . *[My own*
> *"limited powers"] are not suited to that more scientific and more*
> *laborious process of elaborate induction, which in Mr. Darwin's*
> *hands has led to such brilliant results.*

Wallace's interests were not limited to science. In the years after his return to England in 1862, he took on several causes with little regard for how they would affect his reputation among friends and colleagues. Since youth, he had done nothing by halves, and, as noted previously, he seemed not to mind and hardly to notice being considered eccentric or unorthodox. He publicly opposed enforced smallpox vaccination, eugenics, capital punishment, and vivisection, and supported land nationalization and women's education and rights. Not surprisingly, he was particularly concerned about human encroachment on the natural world. He also eloquently argued that if those less fortunate in society were not given greater influence in government and commerce, European civilization would never be superior to the better class of savages he had met during his travels across the globe.

Wallace's esteem as a leading light in scientific circles was tarnished considerably when he engaged in a scientifically rigorous study of spiritualism, particularly of the activities of mediums at séances, and came to the reluctant conclusion that, though some were charlatans, many were not, and spiritualism and its claims were in large part not a hoax. That conclusion cost him dearly. As he wrote,

> My scientific friends are somewhat puzzled to account for what they consider to be my delusion, and believe that it has injuriously affected whatever power I may have once possessed of dealing with the philosophy of Natural History. . . . Facts, however, are stubborn things. My desire for knowledge and love of truth forced me to continue the inquiry. The facts became more and more assured, more and more varied, more and more removed from anything that modern science taught or modern philosophy speculated on. The facts beat me. They compelled me to accept them *as facts* long before I could accept the spiritual explanation for them; there was no place in my fabric of thought into which it could be fitted. By slow degrees a place was made; but it was made, not by any preconceived or theoretical opinions, but by the continuous actions of fact after fact, which could not be got rid of in any other way.

Wallace's reputation recovered. As he grew older, colleagues and the public chose to regard him as a brilliant, distinguished, kindly, modest gentleman who could be forgiven some quirky ideas. To a younger generation of biologists, he was one of the "Grand Old Men of Science." He never gave up his disdain for the idea that findings that didn't fit into the known laws of science should be ignored. It was just there, where something didn't make sense and seemed to undermine all former belief, that true scientific discovery took place. Hadn't that happened with his and Darwin's theory?

Wallace served as a pallbearer at Darwin's ostentatious Westminster Abbey funeral in April 1882, and he outlived Darwin by thirty-one years.

After the attacks it had suffered in the 1860s, the theory of natural selection hadn't encountered another serious challenge for more than a decade; but after Darwin died, opposition came from within the ranks of

An engraving depicting the funeral ceremony of Charles Darwin at Westminster Abbey on April 26, 1882; Wallace served as a pallbearer.

LOST SCIENCE

evolutionists themselves. Wallace fought that battle by himself and won a decisive victory in the early 1890s, ending almost all serious attempts to prove that the theory of natural selection was not the primary explanation for the origin of species. Lesser challenges to the theory—both reasonable and unreasonable—continued to crop up; and into his nineties, Wallace continued to swat them down.

Wallace died at age ninety-one, on November 7, 1913, and was buried in Broadstone. In 1915, a medallion in his honor was placed in the Abbey.

Not everyone has agreed with Wallace that his treatment by Darwin, Lyell, and Hooker in 1858 was fair and honorable. John Langdon Brooks's 1984 book *Just Before the Origin: Alfred Russel Wallace's Theory of Evolution* speculated that Darwin, having in hand both Wallace's essay with the Sarawak Law and his March 1858 essay, altered his own thinking, which he had expressed less clearly in his earlier sketches than he, Lyell, and Hooker claimed. The ideas in his *Origin of Species*, Brooks insists, are Wallace's—*not* anything Darwin had come up with before reading Wallace's essays. In effect, Brooks claims that Darwin plagiarized Wallace.

In Arnold Brackman's 1980 book *A Delicate Arrangement: The Strange Case of Charles Darwin and Alfred Russel Wallace*, the "delicate arrange-ment" refers to Lyell's and Hooker's handling of the presentation to the Linnean Society. Brackman focuses on the class difference between most of the men in the Society and lower-middle-class Wallace, and accuses this "elite" of favoring one of their own. Those present—only a small percentage of the membership—took little notice of either man's work and adjourned unaware that anything dramatic had been proposed. Lyell and Hooker had expected and hoped for this non-reaction and, in a word, contrived it to get the papers in the record with as little notice as possible. Also significant, of course, was the order in which the papers were presented, implying that Wallace's paper merely supported Darwin's ideas.

Wallace surely would have deplored Brooks's and Brackman's insin-uations. He was awarded the Linnean Society's first Darwin–Wallace Gold Medal in 1908—struck that year to celebrate the fiftieth anniversary

174

celebration of the first joint publication of his and Darwin's papers. Wallace was eighty-five years old, and public appearances were rare. In his acceptance speech, he clearly laid out—"setting matters straight," as he put it—why he had never contested Darwin's right to priority:

> *The idea [of natural selection] came to me, as it had come to Darwin, in a sudden flash of insight: it was thought out in a few hours . . . then copied on thin letterhead and sent off to Darwin—all within one week. Such being the actual facts, I should have had no cause for complaint if the respective shares of Darwin and myself had been thenceforth estimated as being, roughly, proportional to the time we had each bestowed upon it when it was thus first given to the world—that is to say, as 20 years' work is to one week. For, had he already made it his theory, after 10 years' work—15 years'—or even 18 years' elaboration of it—I should have had no part in it whatever, and **he** would have been at once recognised, and should ever be recognised, as the sole and undisputed discoverer and patient investigator of the great law of Natural Selection. . . . It was really a singular piece of good luck that gave to me any share whatever in the discovery.*

History has chosen to agree with this humble, gracious man.

The front and back of the Darwin-Wallace medal, first struck July 1, 1908, by the Linnean Society of London.

SEVEN

Lise Meitner

ESCAPE TO OBSCURITY

(1878–1968)

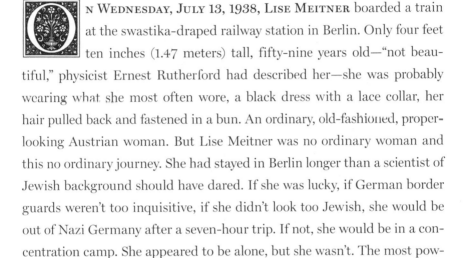

O N WEDNESDAY, JULY 13, 1938, LISE MEITNER boarded a train at the swastika-draped railway station in Berlin. Only four feet ten inches (1.47 meters) tall, fifty-nine years old—"not beautiful," physicist Ernest Rutherford had described her—she was probably wearing what she most often wore, a black dress with a lace collar, her hair pulled back and fastened in a bun. An ordinary, old-fashioned, proper-looking Austrian woman. But Lise Meitner was no ordinary woman and this no ordinary journey. She had stayed in Berlin longer than a scientist of Jewish background should have dared. If she was lucky, if German border guards weren't too inquisitive, if she didn't look too Jewish, she would be out of Nazi Germany after a seven-hour trip. If not, she would be in a concentration camp. She appeared to be alone, but she wasn't. The most powerful British spy in Nazi Germany, Paul Rosbaud, was seeing her onto the train, and Dutch physicist Dirk Coster would travel with her but remain at a distance until shortly before they reached the border. All over the world, friends and scientific colleagues were waiting for the coded signal that Lise Meitner had escaped.

OPPOSITE: *Photograph of Lise Meitner, 1927.* LEFT: *Nazi swastika banners in front of the Berlin Cathedral, 1937.*

Meitner was abandoning everything she cared about: the work of more than thirty years; her beloved laboratory; all her belongings that wouldn't fit into three trunks; and the prospect of a major discovery, just a few weeks ahead, that would have made hers one of the greatest scientific names of the twentieth century. Did she suspect that this frightening, lonely journey—one with no certain future at the other end even if she did get through safely—was a journey into obscurity?

Her life had begun in a different world. She was born in November 1878 into a middle-class family of Jewish extraction in Vienna, the third eldest of eight children. Lise's childhood was a loving, happy one. Her father Philipp Meitner was a lawyer, her mother Hedwig (Skovran) Meitner a musician. Music, reading, and, to an only slightly lesser degree, mathematics and science were at the center of life and interest in the household, but above all, music, a love and frame of reference that would stay with Lise all her life. The words she used much later in her essay "Looking Back" were not inconsequential. She apologized for picking out from the history of physics in the first half of the twentieth century only those things that "I especially remember, and which form as it were a magic musical accompaniment to my life."

A watercolor by Austrian painter Franz Alt of a Viennese street, painted in 1877, a year before Meitner was born in Vienna.

Lise's father was one of the first Jewish men allowed to study and practice law in Austria, and prejudices still existed. Conversion to Christianity was the road to judgeships and the civil service. Families like the Meitners, however, were by no means relegated to the fringes of society. Practicing Jews and those of Jewish ancestry were in large part responsible for making Vienna—the capital of the Austro-Hungarian Empire—also the cultural and intellectual capital of the world. Nevertheless, Lise's parents drew away from the traditions of their own parents, stopped attending synagogue, and had all their children baptized.

The Meitner home was a gathering place where intense conversations about politics, art, literature, and music were the norm. The children sometimes stayed up late for these discussions, and other times didn't. According to her sister Frieda, eight-year-old Lise hid a math book under her pillow and covered the crack beneath the bedroom door so that she could read at night without her parents seeing her light. Not that Lise's parents discouraged independence. Her mother advised her, "Listen to me and to your father, but think for yourself!"

Girls in Vienna received mandatory education until age fourteen, but all eight Meitner children, girls and boys, went on to advanced education. Their parents were realistic people, aware of the difficulties a girl would face at university, and when Lise told her father that she wanted to attend the University of Vienna to study science, he and his wife insisted that she first learn a skill that could support her. Lise earned credentials as a French teacher but continued studying late at night for the University's entrance exams. She had decided on physics, "a real course of study."

Lise was nineteen in 1897 when Austrian universities first opened enrollment to women, but getting accepted was no easy matter. She lacked formal education in the subjects covered in the *Matura* exam* that marked the end of Gymnasium years for young men and was necessary for admission to university: Greek, Latin, mathematics, physics, botany, zoology,

* The *Matura* exam was roughly equivalent to a modern-day American General Education Diploma (GED) exams or British General Certificate of Education Advanced Level (GCE A Level) exams.

A drawing of the main building of the University of Vienna, where Meitner matriculated at age twenty-three.

mineralogy, psychology, logic, religion, German literature, and history. Lise's sister Gisela, similarly unprepared, had proceeded to pack eight years of study into two and win acceptance into medical school. Lise followed Gisela's example. So unremitting and intense were her own two years of study that her siblings teased her "Lise, you are going to flunk, you have just walked through the room without studying!" Fourteen women took the *Matura* exam in July 1901. Four passed. The following October, Lise Meitner sat in lectures at the University of Vienna surrounded by men—the first woman admitted to the physics department. She was twenty-three.

Most of the lectures, labs, demonstrations, and discussion groups for which the over-eager Lise registered took place in a nearly derelict converted apartment building. The entrance reminded her of a henhouse, and she later recalled thinking "If a fire breaks out here, very few of us will get out alive." The teaching was some of the best in the world, but the lectures of physics professor Franz Exner nearly caused her to dose off. Exner was, however, responsible for making Vienna an early center for research into radioactivity, and that was where Lise's future lay.

A 1902 photograph of prominent Austrian physicist Ludwig Boltzmann, whose lectures at the university had a profound influence on Lise Meitner and others.

In her second year, her physics courses—analytical mechanics, electricity and magnetism, elasticity and hydrodynamics, acoustics, optics, thermodynamics, kinetic theory of gases, mathematical physics, and philosophy of science—might seem to have required a host of professors, but they were all taught by one, the eminent theoretical physicist Ludwig Boltzmann. In his lecture room there was no danger of dozing off. Lise Meitner and many others were mesmerized by this brilliant, engaging, self-deprecating man, who told them at the beginning of the course that he was offering them "everything I have: myself, my entire way of thinking and feeling," and asking from them "strict attention, iron discipline, tireless strength of mind . . . [and] that which means most to me: your trust, your affection, your love—in a word, the most you have the power to give, yourself." It was Boltzmann, according to Meitner's nephew, Otto Frisch, who "gave her the vision of physics as a battle for ultimate truth, a vision she never lost."

Boltzmann was convinced of the real existence of atoms and that the atom was divisible, and he had received a great deal of opposition for those views. Ernst Mach had put it bluntly: *"Hab'n S'ein's g'sehn?"*—"'Ave y'seen one of 'em?"

By the time Lise Meitner sat in Boltzmann's lectures in 1902, those disputes were ending in his favor. Radioactivity had been discovered in 1896 and the electron in 1897. Atoms could no longer be thought of as half-imaginary specks of mass that might not exist at all. They were complex structures that could be measured and, potentially, divided, made up of particles with electric charge and loaded with huge amounts of internal energy.

Lise Meitner in Vienna, c. 1906, around the time she received her PhD from the university.

Meitner received her PhD, summa cum laude, from the University of Vienna in February 1906. Her thesis was published in the *Proceedings of the Vienna Academy.* Degree in hand from one of the finest programs in the world, she nevertheless now faced a discouraging situation. There were no professional opportunities in physics open to women. Meitner, however, was not one to be easily deterred, and when Boltzmann, an exuberant man who also suffered periods of deep depression, committed suicide in September 1906, her determination to stay in physics grew all the stronger. She taught for a period at a girls' school, at the same time continuing her scientific education by working in Boltzmann's institute labs at night, learning experimental procedures in the new field of radioactivity. She was increasingly intrigued with that field, but the idea of specializing in it hadn't yet occurred to her.

Having studied with one of the great men of science, Meitner sought out another: Max Planck at the University of Berlin. She went to Berlin in 1907 unaware that women were not allowed to matriculate at Prussian universities and could only audit courses. Nevertheless, she stayed, supported by an allowance from her family. When she asked Planck for permission to attend his lectures, he agreed but said he failed to understand why she wished to study when she already had her PhD. Meitner had also already published original work, but Planck seemed not to have known that, and she was too modest and shy to mention it. Meitner at first interpreted what seemed

Portrait of German physicist and Nobel Prize-winner Max Planck, c. 1930. After auditing his lectures at the University of Berlin, Meitner eventually became his assistant.

an off-putting comment to mean that he had a low opinion of women, but Planck was not, in fact, opposed to women's attendance at university for the very few (as he saw it) with exceptionally extraordinary gifts. Since he later made Meitner his assistant, he clearly came to regard her as one of those gifted few. Planck's lectures seemed dry and arrogant after the warmth and excitement of Boltzmann's, but Meitner soon realized that he was just a very reserved man and that, contrary to her first impression, there was not a trace of arrogance in him. Her description of Planck is one of the best and most delightful we have:

> *His house was the center of good companionship. In the summer we ran races in the garden, and Planck joined in with an almost childlike eagerness and pleasure. . . . Planck once told us that Josef Joachim, with whom he often played chamber music, was such a wonderful man that when he went into a room, the air in the room became better. Exactly the same could be said of Planck.*

Attending lectures left time for experimental work, and Professor Heinrich Rubens, head of the University's Institute of Experimental Physics, offered Meitner a place in his laboratory. She was hesitant. She feared she lacked the courage to admit ignorance and ask Rubens questions about things she didn't understand. Then Rubens introduced her to Otto Hahn. Hahn, a year younger than she, had been a star student of the famous Ernest Rutherford and had already discovered a new element— "radiothorium." He was now in the University's Institute of Chemistry and

Lise Meitner and Otto Hahn in a laboratory at the University of Berlin, 1913.

Biochemistry studying radioactive elements. Like Meitner, he was attending Rubens's physics colloquium every Wednesday, and he liked the idea of collaborating with her. Meitner was comfortable with Hahn and felt she could ask him naïve questions without embarrassment. They would work as colleagues for thirty years.

Women were forbidden entry to the labs of the Institute of Chemistry and Biochemistry (a woman's hair, it was feared, might catch fire, though no one seemed concerned about beards), so in the winter of 1907 Hahn and Meitner set up their lab in a shabby former carpenter's room in the basement. Rutherford, Hahn's former mentor, had only recently discovered alpha and beta rays, and Hahn and Meitner embarked on a project to examine the beta radiation from every known particle. Meitner developed a new way to analyze radioactive processes, known as the "recoil method," which spread to the wider physics community. The Hahn-Meitner team research would soon be recognized well beyond Berlin.

BETA RADIATION

Beta radiation is a radioactive process in which the nucleus of a radioactive atom emits an electron (along with an unusual particle called an *antineutrino*). Because this electron comes from the nucleus of the atom, it is called a *beta particle* to distinguish it from the electrons that orbit the atom.

Most of Meitner's acquaintances were not chemists but other young physicists attracted to Berlin because of Planck's groundbreaking work. These men visited the basement lab often, climbing in through the window rather than using the door, and many would remain lifelong friends. "We were young, contented, and carefree, perhaps politically too carefree." In a piece she wrote many years later, "Looking Back," Meitner would say of these friends, "Not only were they brilliant scientists—five of them later received the Nobel Prize—they were also exceptionally nice people to know. Each was ready to help the other, each welcomed the other's success."

Professors and students met each week for Rubens's physics colloquium. In Meitner's words, it was

> an exceptional intellectual center. All the new results which were then pouring out were presented and discussed there. . . . I remember lectures on astronomy, physics, chemistry—for example a lecture on the stars of various ages given by Schwarzschild, a theoretical astronomer. . . . It was quite extraordinary what one could acquire there in the way of knowledge and learning.

In 1908, Ernest Rutherford received the Nobel Prize in Chemistry. He and his wife, returning from Stockholm to Manchester, England, made a detour to visit Berlin. Rutherford was familiar with Meitner's publications and her recoil method and had been looking forward to meeting her. "Oh, I thought you were a man!" he exclaimed, when they were introduced. While

Ernest Rutherford at McGill University in Montreal, Canada, 1905.

Hahn and Rutherford had long talks about physics and chemistry that Meitner would have relished, it fell to her to take Mrs. Rutherford Christmas shopping. Back home in Manchester, Rutherford wrote, "Lise Mitner [*sic*] is a young lady but not beautiful so I judge Hahn will not fall a victim to the radioactive charm of the lady." Meitner, shy and reserved, and Hahn, self-confident and outgoing, had become close friends. They often sang Johannes Brahms's *lieder* while they worked—Hahn singing beautifully and Lise humming—but they didn't take meals together and until 1923 referred to one another with the formal German "*Sie*" rather than the familiar "*du*."

Meitner lived frugally, still dependent on a stipend from her parents. She rented single rooms without private baths, sometimes with a piano or telephone. She ate little but bought cigarettes and a daily newspaper, and she attended many concerts, sitting up high in the inexpensive rows, often following the music from a full score. For Christmas and often for Easter and summer holidays she returned to her family in Vienna, which still felt like home.

Albert Einstein, c. 1907, approximately age twenty-eight, while he was a clerk at the patent office in Bern, Switzerland.

In 1909, Meitner and Albert Einstein met for the first time when both lectured at a conference in Salzburg, Austria. Einstein was six months younger than she and had, only two months earlier, resigned his job as a clerk in the patent office in Berne, Switzerland. Einstein's first lecture—his first "invited lecture" anywhere—was about developments in understanding the nature of radiation. He proposed that particles of light (now known as "photons") were packets of energy (quanta) and that this was a better way of understanding light than to think of it as waves. He hoped a new theory would view light as a "kind of fusion of wave and corpuscular theories." After the lecture Meitner had to explain to Hahn precisely what Einstein had meant. She predicted that it would have tremendous importance for their work.

In his second Salzburg lecture, Einstein showed how the theory of relativity could be derived from the equation $E=mc^2$, and that an inert mass lay behind every radiation. Years later Meitner would write that these two facts were "so overwhelmingly new and surprising that, to this day, I remember the lecture very well." Though Planck had been one of the first people to realize the significance of Einstein's 1905 theory of special relativity, Meitner hadn't previously paid it much mind, nor did she foresee how it would revolutionize concepts of time and space or that $E=mc^2$ would underlie her most momentous achievement years later. Four years after the Salzburg conference, Planck brought Einstein to work in Berlin, where he joined the group of young physicists not only as a scientist but also as a violinist in chamber music at Planck's house.

Meitner had found the area of science that would engage her for the rest of her life—radioactivity and the field to which it was giving birth: investigating the nucleus of the atom, or "nuclear physics." Nevertheless, there seemed to be no promising career path for her, and she was not entirely free from "negative thoughts." Grieving for her father after his death in 1910, she wrote to a friend

> *Sometimes I lack courage, and then my life, with its great insecurity, the constantly repeated worries, the feeling of being an exception, the absolute aloneness, seems almost unbearable to me. And what distresses me most is the frightful egotism of my current way of life ... everyone should be there for others ... somehow our lives should be connected with others, should be necessary for others. I, however, am free as a bird, because I am of use to no one. Perhaps that is the worst loneliness of all.*

In 1909, women were finally able to enroll at Prussian universities. Meitner was allowed access to the University of Berlin's labs and a newly installed women's toilet. In 1912, she became assistant to Planck in the University's Institute of Theoretical Physics, the first woman to hold a paid position in the University. It was five years since she had arrived in Berlin, and her research was internationally known, yet her duties as Planck's "assistant" included correcting his students' assignments. Nevertheless, Meitner later called this appointment the turning point of her professional life, giving her a "passport to scientific activity in the eyes of most scientists and a great help in overcoming many current prejudices against academic women." Her published research results were winning respect in her own physics circle and in the international physics community. In 1908, she and Otto Hahn published three major articles, and in 1909, six. In 1910–1913 they published fourteen. It frequently fell to Meitner to be the one to interpret their results. Rutherford once wrote of Hahn, "You seem to have a special smell for discovering new radioactive elements," but it was often the more keenly perceptive Meitner who recognized what they were smelling.

An engraving depicting Elsa Neumann's graduation from the University of Berlin in 1899. Neumann, who had to individually petition professors to attend their lectures, was the first woman to receive a doctoral degree in physics from the university.

She was an important participant in an emerging field, and the Berlin physics and chemistry hierarchy were obliged to treat her as such. She became a Scientific Associate in the University's new Kaiser Wilhelm Institute for Physical Chemistry and an unpaid "guest physicist" in Hahn's radioactivity department lab at the Kaiser Wilhelm Institute for Chemistry, both newly built in Dahlem, a suburb of Berlin. Laboratorium Hahn-Meitner, the joint radioactivity section, was truly theirs and housed in clean, spacious rooms where, for the first time, measurements were taken to avoid chemical and radioactive contamination. Meitner gave a splendid hotel dinner party to celebrate. However, at the weekly physics colloquium, she still always allowed Hahn to be the one to report on their joint research results and papers.

Rutherford had introduced a new model of the atom. It is sobering to realize that it wasn't until the period between 1911 and 1914 that what has become the more popular mental picture of an atom appeared—a miniature solar system with electrons like planets orbiting the nucleus—replacing the older picture of positively charged parts of an atom spread out through the whole like raisins in raisin bread. As soon as Rutherford

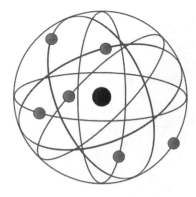

The Rutherford-Bohr model of the atom.

announced the new model, young Niels Bohr from Copenhagen, who had helped Rutherford arrive at it, decided it was unsatisfactory. Bohr placed the electrons in specific orbits. In what has come to be known as the Rutherford-Bohr model of the atom, electrons with negative charge orbit a nucleus with positive charge in orbits that have set sizes and energy. The lowest energy is found in the orbit closest to the nucleus. When an electron absorbs energy, it moves to an orbit farther from the nucleus. When it emits energy it moves to an orbit closer to the nucleus. Because energy is "quantized," there are no "in between" orbits. Bohr's quantized atomic structure set the world physics community afire with discussion and controversy, with younger physicists in particular, including Meitner, supporting it, and Einstein exclaiming "It is one of the greatest discoveries!" Many agreed that even if this model needed refining, it was a truly magnificent "attempt" (as Bohr himself called it).

Meitner and Hahn began a fresh project: the search for the immediate precursor, the mother substance, of the element actinium*—now number 89 in the periodic table of elements. Put more simply, the question was: What substance, through radioactive decay, produces actinium? Pure uranium had never been found to generate it, so it was suspected that the two must have an "intermediate." Meitner and Hahn defined their task as "to find that substance which forms the starting point for the actinium series,

---

* Actinium: A soft, silver-white, strongly radioactive, metallic element found in uranium ore. It glows in the dark with a pale blue light. Its atomic number (the number of protons in an atom's nucleus) is 890. Its atomic symbol is Ac.

and to determine whether and through which intermediate it is derived." That work was interrupted by the beginning of World War I.

By the end of 1914, Hahn had been called up to serve in the German army. Meitner continued searching for the actinium precursor but widened her research into medical applications of X-rays and trained as an X-ray technician/nurse. The head of the Institute for Physical Chemistry, Fritz Haber, was devoting himself and the institutes at Dahlem to the study and development of defensive and offensive measures in gas and chemical warfare. Meitner was pleased when Hahn was reassigned to work at Haber's division. They could, to some extent, continue their collaboration. But she was also deeply concerned about the secret research in which he was involved, suspecting what it was. She had strong reservations about the morality of chemical warfare.

In the summer of 1915, Meitner volunteered as a nurse/technician and was assigned to a hospital near the Russian front in one of the first hospital X-ray units. Experiencing the worst of front-line nursing, she wrote to Hahn "That there is physics, that I used to work in physics, and that I will again, seems as much out of reach as if it had never happened and never will again in the future . . . [but] during the day, I think only about my patients." After a move to Poland to a hospital so ill-equipped and exhausted that it was incapable of treating many wounded, Meitner returned to Berlin in October 1916. The institutes at Dahlem were almost entirely involved in war research, but Meitner was able to keep Haber from taking over the section that had been hers and Hahn's. Hahn was in Galicia in Eastern Europe, testing new chemical warfare inventions at the Russian front. In March 1918, when Meitner discovered the substance whose radioactive decay produces actinium—protactinium—Hahn was still away, but she gave him full credit as first author rather than herself.

By the time the war ended, physics had been transformed, largely through Niels Bohr's work on the structure of the atom. Chemists and physicists were at last able to correctly explain the periodic system of the chemical elements. The Institute, with private support, was officially divided into two. Meitner was asked to establish a Department of Radioactive Physics and serve as its director. Hahn continued in the Chemistry Division. After about 1920, they no longer worked together directly, though the work of each continued to complement the other.

A serious difficulty for German scientists after the war was that they were excluded from all scientific conferences. Bohr, in Sweden, strove to end this exclusion, and it was an extraordinary privilege and exception to be invited to give a lecture in Copenhagen in 1921 on beta and gamma radiation. There, Meitner got to know Bohr and his wife on a close, personal basis for the first time. That same year Meitner and her assistant Kurt Philipp were the first to identify positrons (the positively charged antiparticles of electrons) from a non-cosmic source and to show that they appear in pairs with electrons, which are negatively charged.

In 1922, Meitner received "Habilitation" at the University of Berlin, a promotion that finally gave her the right to lecture and teach, and became a "Privatdozentin"—one of the first women in Germany to achieve this status. The title carried no stipend but allowed one to take paying private students. The newspaper published her inaugural lecture "The Problems of Cosmic Physics" under the title "The Problems of Cosmetic Physics."

Otto Hahn and Lise Meitner in a lab in 1919, the last year they would work together side by side until 1934.

The years between the wars were extremely productive for Meitner. She kept closely in touch with the work at Bohr's institute in Copenhagen and leading scientific institutions worldwide. She worked on X, beta, and gamma radiation and the relationships among them, often not stopping until late at night to return to her apartment. She proved that after beta decay, there is secondary beta radiation—gamma radiation—from the electron shell of the *new* atom, not from the original atom. Meitner's research and papers—ten published each year in the mid-1920s—became known all over Europe and in America. Well-deserved and highly gratifying honors came her way: In 1924, she was the first to receive the American Association to Aid Women in Science Award; and she was the first woman to receive the second-place silver medal of the Leibniz Prize. Each year for ten years, the team of Hahn and Meitner were nominated for the Nobel Prize—in six of those years by Max Planck.

The possibility of the "release of atomic energy" was being widely discussed. Might not atomic power someday light homes and factories, and power submarines or even spacecraft? Or was that all science fiction? In 1932, James Chadwick discovered the neutron—the "bullet" that would pierce the atomic nucleus to its heart—but in 1933, Einstein expressed the opinion that attempts at "loosening the energy of the atom [are] fruitless." Rutherford scoffed that "anyone who expects a source of power from the transformation of atoms is talking moonshine." Meitner thought otherwise.

With Hitler's rise to power, Max Planck and friends and scientific colleagues around the world began increasingly to fear for Einstein and Meitner. Both, along with hundreds of others of Jewish descent—professors, researchers, and other academics—were dismissed from their German university positions. Though her title as professor was revoked in April 1933, Meitner hung on.

She and Hahn, working directly together again in 1934 for the first time in twelve years, applied themselves to following up on Italian Enrico Fermi's recognition that because neutrons lack charge, they are "suited to penetrate heavy elements—that is, elements high on the periodic table— and of how suitable they were to the release of nuclear reactions." They observed the strange behavior of atomic nuclei under bombardments

A portrait of scientists attending the Seventh Solvay Conference in Brussels, October 1933, organized by the International Solvay Institutes for Physics and Chemistry. Meitner is seated second from the right. Other illustrious attendees included Niels Bohr, seated third from the left; Marie Curie, seated fifth from the left; and Ernest Rutherford, seated sixth from the right. Meitner's title as professor had been revoked six months earlier.

with slow neutrons, which resulted in the production of what might be transuranic elements—elements beyond uranium in the periodic table of elements. They were one of the few teams of researchers in the world prepared to deal with the chemical and physical difficulties this involved. Hahn reported that "what happened when uranium was bombarded with neutrons was very complicated indeed."

In March 1938, Germany annexed Austria. It was becoming increasingly dangerous for Meitner to remain in Berlin and increasingly "embarrassing" for the Institute to keep her on. The head of a new Guest Department of the Institute for Chemistry, a member of the Nazi party, remarked: "The Jewess endangers the Institute." Hahn hurt Meitner deeply when he failed to defend her in a conversation regarding the "awkward" situation her association with the Institute was creating. He discouraged her from continuing, even on an unofficial basis. "Hahn says I should not come to the Institute any more," she scribbled in her diary. When she went there anyway because there were irradiation results that needed recording, she wrote, "He has actually thrown me out."

Very few German colleagues were brave or foolhardy enough to defend her, but Carl Bosch, president of the Kaiser Wilhelm Society, rejected the idea that Meitner resign. He knew she needed to be employed in order to have permission to leave Germany. Friends and colleagues abroad offered invitations to lecture, faculty positions, and financial support. In early May, after receiving an invitation from Niels Bohr to come to his institute in Copenhagen, she finally decided to abandon her beloved, highly promising work in Berlin. It was a heart-wrenching decision and one that eventually cost her the Nobel Prize, but it saved her life.

FRONT ROW L-TO-R: *Niels Bohr, Werner Heisenberg, Wolfgang Pauli, Otto Stern, Lise Meitner, and Rudolf Ladenburg in an auditorium with other physicists, possibly in Copenhagen for a Nobel Prize lecture, c. 1937. Meitner would soon be faced with the need to flee Berlin.*

Departure was no simple matter. The Danish consulate refused a travel visa, because her Austrian passport was no longer valid after the German annexation of Austria. The decision whether to *leave* Germany had become a different question: how to *escape*.

Bosch contrived to help her apply for a permit to allow "the famous scientist Lise Meitner" to travel out of Germany to a neutral foreign country, cleverly framing his request in terms of documentation that would allow Meitner to *return* to Germany and even ending his letter with "Heil Hitler!" The ploy was unsuccessful. Bosch read the reply over the phone to Meitner, who for safety's sake had moved to a hotel room across the street from the Institute: "It is considered undesirable that renowned Jews should leave Germany for abroad to act there against the interests of Germany." The word Jews was underlined in the message.

Though her friendship with Hahn was strained—and in spite of his discouraging her from doing so—she went every day to the Institute and urged Hahn to continue bombarding uranium with neutrons. During the first few days of July 1938, she spent long hours in the labs with him and

Hitler accepts the salutes of the Reichstag in March 1938 after announcing the "peaceful" annexation of Austria by Germany.

their primary assistant Fritz Strassmann, discussing their discoveries and publications. Then, on July 4, Hahn learned from Bosch that the prohibition against scientists leaving Germany was about to be more strictly enforced. It was urgent that she get out.

Two decades earlier, Meitner had met Dutch physicist Dirk Coster and had become friends with him and his wife. It was Coster who, at this supremely critical juncture, traveled to the border town of Nieuweschans, met personally with immigration officers, showed them Meitner's entrance permit from the Hague, Netherlands, and requested that they use "friendly persuasion" with the German border guards to let her cross into the Netherlands. Then Coster boarded a train swarming with German soldiers, headed for Berlin. The journey was a nightmare. At the border, passports were repeatedly checked before the train was allowed to continue. Out his train window, Coster saw that every little town had swastika flags hanging from its medieval buildings. It seemed impossible that he could bring Meitner out.

Hahn had told her the plan. They would continue to work through the day at the Institute as though nothing out of the ordinary were happening. Then they would go to her hotel, meet Paul Rosbaud—an old friend who was editor of the journal *Die Naturwissenschaften*, physics consultant for the publisher Springer Verlag, *and* Britain's most valuable spy in Nazi Germany—and pack her belongings. Meitner later gratefully remembered how Rosbaud helped, "when you, with great friendly understanding, went through my rooms and put everything possible into my trunks." Rosbaud drove her to Hahn's house to spend the night. Hahn gave Meitner a diamond ring to use for a bribe or to sell in a financial emergency. It had been his mother's. On the morning of Wednesday, July 13, Rosbaud and Hahn returned to her hotel and packed one more suitcase for her, giving no indication that she was checking out. Everyone was to think she had gone on a summer holiday. Rosbaud drove her to the station. Meitner was tense and fearful, and it wasn't easy for him to persuade her to board the train. Coster was on the same train, but the plan was that they would meet up, and then very discreetly, only as they neared the border.

Meitner's friends held their breath:

> We had agreed on a code telegram by which we would be let know whether the journey ended in success or failure. The danger consisted in the SS's repeated passport control of trains crossing the frontier. People trying to leave Germany were always being arrested on the train and brought back.

It took seven hours to reach the border. Coster joined Meitner where she was seated. They watched as uniformed German officers worked their way nearer, scrutinizing documents. The officers approached, and the Dutch guards peered at her over the shoulders of the German officers. She later admitted that she nearly fainted, so great was her fear and tension—like nothing she had ever experienced—but the officers and guards passed on. At last the train moved forward. Several hours later Meitner and Coster alighted in the university town of Groningen. Coster picked up her bags and took a shaken and exhausted Meitner to his car. The next day he wired Hahn that the "baby" had arrived, and Hahn relayed the news to friends in Berlin and across the world. Years later, Coster's widow commented that she wondered whether the German border guards might have ignored "Frau Professor," thinking that she was Coster's wife. If so, then for once it was an advantage to be a woman.

Meitner was no longer young; she would be sixty in November. Nuclear physicist Adriaan Fokker observed that she seemed "inwardly torn apart" as he and others joined the Coster family in trying to make her feel welcome without crowding her. After about a week of total exhaustion, Meitner began to explore the lovely university town of Groningen, where there were no swastikas, no blaring Nazi voices on loudspeakers, no goose-stepping soldiers. Coster received a barrage of congratulatory telegrams and letters for his part in "the abduction of Lise Meitner."

In late July, Coster and Meitner discussed with Bohr the possibility of Meitner's joining Bohr's Swedish Research Institute for Physics, in

Niels Bohr, c. 1930s.

Stockholm, which was just completing construction. Meitner traveled first to Copenhagen, where the Bohr family welcomed her with flowers in her room at the mansion Bohr had been given when he received his Nobel Prize. There, in a house alive with the Bohrs' children and grandchildren, where conversation ranged from family to physics, Meitner, though still "miserable," began to think seriously about her future. In late August, she took the ferry across the Øresund to Sweden, heading for Stockholm and Bohr's institute. Her letter to Hahn announcing her decision to retire formally from her position in Berlin ended with the words "I have, in my inner self, still not entirely realized that this that I have written here is real, but it *is* real."

That fall Meitner corresponded frequently in long letters with Hahn. She told him of her loneliness and frustration trying to work either in her own small room or in labs still under construction where no one wanted to speak German. She was learning Swedish as quickly as possible. In early November she took the eight-hour train ride across Sweden back to Copenhagen to stay again in Bohr's home when Hahn was visiting. With Bohr, she, Hahn, and her nephew Otto Frisch—a younger Austrian nuclear physicist who was working at Copenhagen with Bohr—discussed results that had Hahn puzzled and stymied, and they debated about how those results might be interpreted.

In Berlin, Hahn and Strassman were continuing to bombard uranium with slow neutrons, expecting to produce radium isotopes. The nuclei of most elements changed somewhat during neutron bombardment, but Hahn's uranium nuclei were changing much more dramatically, splitting into two roughly equal pieces—not new transuranic elements but isotopes that behaved like radioactive barium isotopes. Hahn thought they might actually *be* barium. Barium was not even near being a transuranic—with the atomic number 56, it was halfway down the periodic table. "A dreadful conclusion," he wrote to Meitner in Sweden on December 19, 1938, "perhaps you can suggest some fantastic explanation. These are all very tricky experiments. But we must clear this thing up. We don't believe this is nonsense, or that contaminations are playing a joke on us."

On a Christmas skiing vacation, Meitner and her nephew Otto Frisch thought about Hahn's problem. Frisch described the occasion in an article and in his memoirs. The date was December 24, 1938:

LEFT TO RIGHT: *Physicists William Penney, Otto Frisch (Meitner's nephew), Rudolf Peierls, and John Cockcroft wearing the American Medal of Freedom for their contributions to the Manhattan Project during World War II, 1946.*

> *I found her at breakfast brooding over a letter from Hahn. . . . We walked up and down in the snow, I on skis and she on foot, and gradually the idea took shape that this was no chipping or cracking of the nucleus but rather a process to be explained by Bohr's idea that the nucleus is like a liquid drop; such a drop might elongate and divide itself.*
>
> *There were strong forces that would resist such a process, just as the surface tension of an ordinary liquid drop resists its division into two smaller ones. But nuclei differed from ordinary drops in one important way: they were electrically charged, and this was known to diminish the effect of the surface tension.*

The two of them sat down on a tree trunk and (as Frisch later described it to colleagues at the Cavendish Labs in Cambridge) began to calculate with sticks in the snow.

I worked out the way the electric charge of the nucleus would diminish the surface tension. . . . Lise Meitner worked out the energies that would be available from the mass defect in such a break-up.

After separation, the two drops would be driven apart by their mutual electric repulsion and would acquire very large energy, about 200 MeV in all. Where could that energy come from? [Meitner] worked out that the two nuclei formed by the division of a uranium nucleus together would be lighter than the original uranium nucleus by about one-fifth of the mass of a proton. Now whenever mass disappears, energy is created, according to Einstein's formula E=mc²; and one fifth of a proton mass was just equivalent to 200 MeV. So here was the source for that energy; it all fitted!

The atom had been split!

In 1950, in his book, *Out of My Later Years*, Einstein would write about Meitner's essential part in the discovery:

I do not consider myself the father of the release of atomic energy. My part in it was quite indirect. I did not, in fact, foresee that it would be released in my time. I believed only that it was theoretically possible. . . . It was discovered by Hahn in Berlin, and he himself misinterpreted what he discovered. It was Lise Meitner who provided the correct interpretation.

Meitner didn't only provide the interpretation. The experiments in Berlin that culminated in the discovery had been hers and Hahn's joint work until she left Germany five months earlier, and Hahn and Strassman had continued with her encouragement and input. The correspondence back and forth had been prolific.

Frisch hurried to Copenhagen to share the news with Bohr, who was about to depart for America. Bohr urged Frisch to write up the discovery immediately (which he did in abbreviated form) because once he reached

America, the cat would be out of the bag. Bohr was concerned that proper credit be given, and his worries were well founded. Astounded researchers, hearing the news at a meeting in Washington (about another subject entirely) rushed out of the room to perform their own experiments. Before Meitner's and Frisch's formal publication in February, the news was already spreading like wildfire in America, causing a sensation among scientists and soon among the press and public. Bohr urged American scientists not to publicize their own findings based on the discovery before Frisch and Meitner published theirs. Frisch and Meitner finally did, in the February 11, 1939, issue of *Nature*, where the word "fission," coined by Frisch, appeared for the first time.

The equipment used by Otto Hahn, Lise Meitner, and Fritz Strassmann in their nuclear fission experiments in 1938 in Berlin.

Unfortunately for Meitner, nuclear fission quickly became known as "Hahn's discovery." In Nazi Germany, Hahn was finding it politically advisable for his own safety and that of colleagues at the institute to disassociate himself from Meitner. If the breakthrough of nuclear fission were his alone, he could hope to be invulnerable both politically and professionally. Not so if he shared recognition with a Jewish woman. His insistence on sole credit was a matter of survival. However, Hahn's attempts to exclude Meitner caused a rift between them that would never fully heal.

In the late winter of 1938, Meitner went on carrying out experiments in Copenhagen that involved collecting fission fragments and testing them for the presence of transuranics. The results were negative. Meitner wrote to Hahn twice on the same day, March 10, reporting those findings. She used the opportunity to berate him only gently for his treatment of her, reminding him that she should receive credit for nuclear fission as well as he: "You and Strassmann could never have made your beautiful discovery without the earlier uranium work." A gentle reproof indeed. She referred to it as "your" discovery.

Among the flurry of articles on fission and fission products that appeared that winter, one was especially significant for Meitner and Frisch: "The Mechanism of Nuclear Fission" by Bohr and American John Archibald Wheeler. The piece would become a physics classic. It caused an international debate about the possibility of chain reactions from nuclear fission and gave Meitner and Frisch priority credit. However, Meitner's discouragement (she hadn't yet been able to bring her belongings out of Germany and truly settle in Copenhagen) sank to a new low when the Nobel Committee decided not to award a Prize in Chemistry in 1939, nor a Prize in Physics the following year. It seemed that her accomplishments (and Hahn's as well) were being ignored. After no fewer than fifteen nominations over the years and one of the greatest breakthroughs of all times, they had been passed over again.

Meitner was increasingly unhappy that nuclear fission was becoming huge science with attempts to realize its potential as a weapon. She continued her own science on a smaller scale and published her results and her own new theories about fission and gamma rays. Still a robust, healthy woman in her fifties, she walked six to eight miles a day. She also served as a conduit for communications between Rosbaud in Germany and his wife in England, corresponding with each and passing one's messages to the other in veiled, coded language. How much she became involved in espionage activities beyond this remains the subject of inconclusive hints—destined to remain an untold story.

It was not until the bombings of Hiroshima and Nagasaki in Japan in August 1945 that Meitner's life changed again. She was stunned, speechless, in tears, and fighting off press intent on celebrating her "work on the atomic bomb." There were good moments: a lengthy transatlantic telephone interview with Eleanor Roosevelt. The two expressed their belief in the dire need for world cooperation, especially with the involvement of women, in the creation of long-lasting peace. Meitner voiced her hope that nuclear energy would be used for peacetime work.

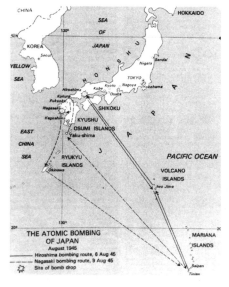

A map illustrating the routes of the U.S. atomic bombing missions of August 1945 in Japan.

Though some in the press dubbed her "the Mother of the Atom Bomb," Meitner continued to insist "I will have nothing to do with a bomb!" In an interview with the *Saturday Evening Post* in 1946, she said: "I myself have not in any way worked on the smashing of the atom with the idea of producing death-dealing weapons. You must not blame us scientists for the use to which war technicians have put our discoveries." Otto Frisch was assigned to work at the top-secret Los Alamos weapons laboratory. Meitner was not.

Many in the international science community and among Meitner's closest friends were shocked when Otto Hahn alone received the 1944 Nobel Prize in Chemistry for the discovery of nuclear fission. The decision was not unanimous and there was serious debate, but it was Hahn who stood on the podium. Meitner was in the audience. In his acceptance speech Hahn mentioned her and Otto Frisch, as well as his assistant Fritz Strassmann, but stopped short of commenting that all should, by rights, have been included in the Prize. Afterward he gave an undisclosed amount of his award money to Meitner, which she gave to the Aid Committee for Atomic Physicists at Princeton, where Einstein was working. Meitner's failure to win the 1944 Nobel Prize in Chemistry was not the end of her disappointments. The 1944 Prize in Physics, for which she was also nominated, went to I. I. Rabi, who had worked closely with J. Robert Oppenheimer at Los Alamos on the atomic bomb. The 1945 Prize in Physics went to Wolfgang Pauli for contribution to the understanding of nuclear structure and discovering the neutrino. Nobel Prizes seemed to fall all around Meitner but always miss her.

In 1946, she accepted an invitation to lecture in Washington, D.C., at Catholic University—all expenses covered—and spent several months in America. At the Women's National Press Club annual awards ceremony, she dined with President Truman, artist Georgia O'Keeffe, and dancer/choreographer Agnes de Mille. Her lecture tour took her to Harvard, MIT, Princeton, Wellesley, and many other colleges and universities where she was awarded honorary doctorates. Wearing her usual black dresses with lace collars, with her hair in a bun, she did indeed look like an old-fashioned,

Meitner meets with finalists at a science talent search competition at the Catholic University in Washington, D.C., where she was invited to lecture in 1946.

proper Austrian woman, but she was a scientific celebrity—second to no male scientist in the world—and she seized the moment to press for opportunities and equality for women in science. Back in Scandinavia, the Royal Swedish Academy of Sciences elected her a foreign member. Only two other women had been so honored in the Academy's two hundred years.

Meitner stayed in Sweden for twenty years as a research professor at the University College of Stockholm. She always retained her Austrian citizenship. Twice she turned down invitations from the institute in Berlin (its name changed to the Max Planck Institute for Chemistry), but she did visit Germany to be awarded—jointly with Otto Hahn—the 1949 Max Planck Medal, the highest honor in German physics.

In a rare display of indignation, Meitner wrote in 1953 to Hahn, then president of the Max Planck Society, expressing her chagrin at having her achievements all but erased in the Society's publications:

> In the report of the Society, the lecture is mentioned that I had given in Berlin (it was purely a physics lecture). In this report, I am called "the long-time coworker of our President." At the same time, I have read an article written by Heisenberg in Naturwissenschaftliche Rundschau concerning the relations between physics and chemistry in the last 75 years. In this article, one single mention was made of me. I quote: "the long-term coworker of Hahn, Fraülein Meitner." In the year 1917, I was officially entrusted by the Governing board of the Kaiser Wilhelm Institute for Chemistry with the development of the Physics Section, and I led that for 21 years. Now, try to understand this from my perspective! Should it be possible during the last fifteen years—a time I would not wish on any good friend—that even my scientific past should be taken away? Is this fair? And why is it like this? What would you say if you were to be characterized as a "long-term coworker" of mine?

By this time she had published more than 105 monographs, books and articles, but Hahn never set the record straight.

In the years after the war, strong efforts to bring Meitner to Britain were vetoed by an adviser to Churchill, Frederick Lindemann, Lord Cherwell, who was outspoken in his aversion to women, Jews, colored people, his mother, his brother, and animal protein. A fall and broken hip in the late 1950s, however, persuaded Meitner to move to Cambridge, England, near Otto Frisch and his family. In 1963, when she was eighty-four, she traveled to Vienna to speak about her lifetime in physics before a large audience, a speech later printed as the retrospective "Looking Back." On a 1964 trip to spend the Christmas with siblings who were living in America, she suffered a heart attack. She returned, weakened, to England and survived long enough to receive, in 1966, the coveted Enrico Fermi Award, along with Hahn, Frisch, and Strassmann. She was the first woman to be honored with that award. At last, someone got the credits right. Meitner died in Cambridge on October 27, 1968, eleven days before her ninetieth birthday.

Meitner's name was used, in 1992, for a newly discovered element, number 109 on the periodic table—meitnerium. Other than that tribute, she largely disappeared from public memory.

Lise Meitner receiving the Enrico Fermi Award from U.S. Atomic Energy Commission chairman Glenn Seaborg. The AEC and President Lyndon Johnson jointly awarded Meitner, Hahn, and Strassman in 1966.

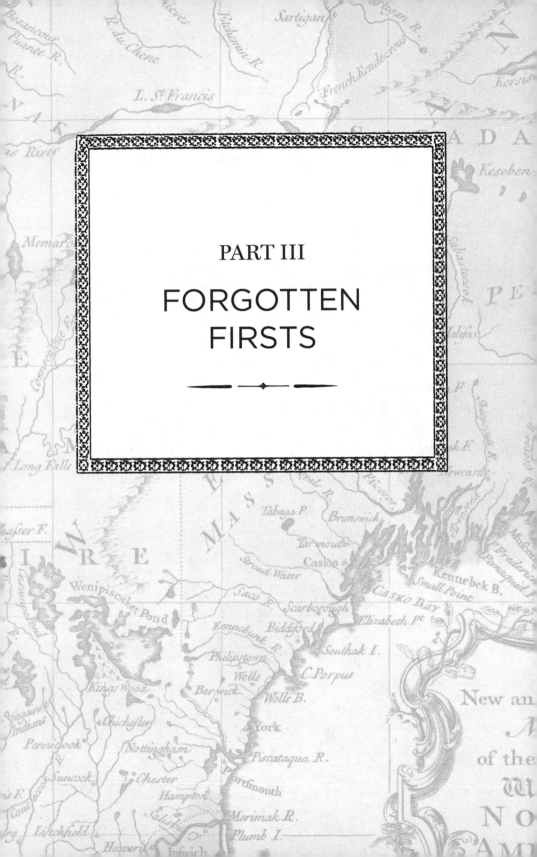

PART III

FORGOTTEN FIRSTS

EIGHT

Johannes Kepler

NEAR-FATAL FICTION

(c. 1590–1634)

 N EUROPE IN THE EARLY SEVENTEENTH century, it was an extremely incautious move to write a story about a skinny crone who could communicate with *daimons* and commandeer them to take her to the moon—if your mother's neighbors thought she was precisely that sort of old woman and were looking for reasons to accuse her of witchcraft. Johannes Kepler's *Somnium* was never intended to reach Katharina Kepler's neighbors. It was an allegorical and scientific book, meant to amuse and inform an erudite audience. Very few of that audience ever read it. In fact, few people of any sort have ever read it, in spite of the fact that it was the first real science fiction book. Unfortunately for Kepler and his mother, the wrong people did read it.

Kepler's *Somnium* has two competitors for "first science fiction book." In the first century CE, Plutarch wrote *On the Face in the Moon*, a tale of a Greek adventurer who discovers an island where the inhabitants know of a passage to the moon. Plutarch

OPPOSITE: *Detail of a nineteenth-century engraving of Johannes Kepler casting the horoscope of Emperor Rudolf II (not pictured), Holy Roman Emperor, in Prague.* LEFT: *Title page of Keppler's Somnium, 1634.*

speculated at length about the possibility of life there. However, there is almost nothing scientifically based in Plutarch's story. The second contender appeared around 165 CE, written by Lucian of Samosata. In his story, an enormous storm carries a Roman galley into the air and all the way to the moon. Fiction this certainly was, though Lucian titled it *A True Story*. Again, there was no science in it. It was a satirical work, a parody of Homer's *Odyssey*.

Somnium, or, to give it its full title, *Somnium seu Astronomia Lunari* (*Dream or Astronomy of the Moon*) is a far more deserving choice for "first science fiction book," written by the man whose three laws of planetary motion still lie at the heart of our understanding of the solar system and other planetary systems. *Somnium* is certainly fiction, with the difference that the science is front, center, and virtuoso. The completed book was a product of Kepler's later years, but it had its beginnings when he was a student, and the book grew and matured as he did. Some of the astronomy that underlies the completed story was unknown to Kepler and everyone else when he first thought and wrote about the moon. But as controversial as Kepler's science might have been in some quarters, it was not the science that made what began as a student "disputation"* a threat to his life and his family a half century later.

In the late 1500s, the University of Tübingen still officially taught the Earth-centered astronomy of Ptolemy, whose ancient vision of the cosmos (rooted in Aristotle) had dominated Western culture and the culture of the Islamic world for fifteen centuries. Kepler's mathematics teacher and mentor at Tübingen (and later his lifelong friend) was Michael Mästlin, one of a mere handful of scholars in all Europe who believed that Copernicus's sun-centered system of the cosmos—which Copernicus had only recently introduced in 1543 in his book *De Revolutionibus Orbium Coelestium*— should be taken literally, and not merely as a mathematically useful hypothesis. Kepler, "partly out of Mästlin's lectures, partly out of myself—collected all the mathematical advantages which Copernicus has over Ptolemy" and came to agree with Mästlin.

---

* A disputation in medieval universities was an academic exercise in which one participant defended a thesis or idea using formal logic and inviting debate from other scholars.

A panoramic map of Tübingen by Swiss engraver Caspar Merian, c. 1650. The university, where Kepler studied, is located to the left of the church spire.

At the time, however, the young Kepler had no plans for a scholarly lifetime in mathematics or astronomy. He was looking forward to a career in theology that would, if circumstances had not intervened, or (as Kepler would have it) had not God intended otherwise, eventually have brought him to a Lutheran pulpit. Even his Copernican leanings had a religious side to them, for wasn't it far more appropriate that the sun, that brightest and most splendid source of light and warmth, should be the center of all things and symbolize their creator? As a university student, he was already the exuberantly religious man that he would remain all his life.

When the nineteen-year-old Kepler received his master's degree in August 1591, the University senate wrote of him: "Young Kepler has such an extraordinary and splendid intellect that something special can be expected from him." They didn't say in what field that might be. Kepler had been carefully fulfilling all the expectations of a theology student, keeping to himself how seriously he rankled at the ugly, un-Christlike disputes between Lutherans and Calvinists. Meanwhile, he was also busily and happily laboring away on astronomical questions, engaging in debates and writing notes for disputations that argued for Copernican sun-centered astronomy. In one of these disputations, written in 1593, Kepler tried to demonstrate that the Earth, far from standing still, was moving in two ways at the same time. It was revolving on its axis *and* was traveling around the sun. He made his argument by calculating how those movements and the movements of other planets would look to moon dwellers, who of course would be no more able to detect or discern the true movements of their home world among the heavenly bodies than we are able to do for our Earth. A friend and fellow student Christopher Besold enthusiastically formulated twenty theses from Kepler's paper and presented them to Professor Veit Müller, the man in charge of arranging philosophical debates at the university. Müller caught too strong a whiff of "Copernican" in Kepler's disputation and Besold's theses and refused to approve a debate. But Kepler had also come across Lucian of Samosata's *True Story*, and the seeds of *Somnium* had been sown. Kepler still had his student paper at hand when he was an old man composing the explanatory notes to *Somnium*.

Two years later, in 1595, after Kepler at the decision of the University of Tübingen had unexpectedly and begrudgingly moved to Graz, Austria, to be a mathematics teacher rather than a theologian or clergyman, he read Plutarch's *On the Face in the Moon*. Plutarch of course lived many years before Copernicus, but his little story showed that he was aware of some of the problems of explaining planetary movements using the ancient astronomy of Aristotle. Kepler was pleased to discover a second classical precedent for his rejected disputation.

Kepler's career and life in the fifteen years after moving to Graz in 1594 make for one of the most astonishing, unlikely tales in the history of science, and unfortunately there is only enough space here to fast-forward through it rapidly. In 1595, while drawing a diagram on the chalkboard for his students, he hit upon an astounding, outside-the-box—and wrong—theory of the arrangement of the planets. He wrote about it in his first

A drawing illustrating Kepler's polyhedral theory, from his first book, Mysterium Cosmographicum *(The Cosmographic Mystery), 1597, shows his nested arrangement of the planetary orbits and the five Platonic solids.*

book, *Mysterium Cosmographicum*, which appeared in 1597. He married in Graz, mourned the death of two infant children, endured persecution and banishment from the city during the Counter-Reformation, and was drawn by convoluted circumstances into a dangerous and dysfunctional but inestimably valuable collaboration with the great astronomer Tycho Brahe. After serving several years as Tycho's rather ill-used assistant in Prague, Kepler survived the mercurial old man and became imperial mathematician of the Holy Roman Empire. It was quite a ride, and quite a rise in the world, and it didn't stop there.

Working with Tycho's splendidly accurate observations of the planet Mars, employing his own mathematical genius, and sticking to an unshakable belief that God had created a harmonious and ordered universe, Kepler discovered that planetary orbits are not circular but elliptical and that the speed a planet travels in its orbit varies in a predictable way, speeding up as the planet comes nearer the sun and slowing as it moves farther away.

215

Kepler chronicled the intellectual adventures that led him to those discoveries (celebrated as the first two of his three laws of planetary motion) in his book *Astronomia Nova*, one of the great watersheds in the history of knowledge. He was now admirably prepared to write his little science fiction–fantasy.

In the summer of 1609, Kepler was well situated in Prague, the capital of the Empire. Although he was seldom paid his promised salary from Emperor Rudolf II, he was more respected and comfortably settled than he had been for most of his life so far. That summer he took great pleasure in engaging in learned and often whimsical discussions with one of his best friends, an ecclesiastical adviser to the emperor who rejoiced in the memorable name Wacker von Wackenfels. In these conversations, *Somnium* reemerged. Emperor Rudolf had asked Kepler to comment on an idea having to do with the patterns of light and shadow on the face of the moon. Rudolf proposed, loosely following Aristotle, that they were reflections of the land masses on Earth. Kepler was obliged to discourage Rudolf. He had known since he was a student and from subsequent years of study and observation that the patterns are shadows caused by mountains and other natural features on the moon itself. Kepler and von Wackenfels discussed the subject repeatedly, and Kepler decided he had enough material and ideas to write a moon geography, following up on his student disputation in a little fable to amuse his friend.

BELOW: *A panoramic map by noted etcher Wenceslaus Hollar of Prague of his native city, 1636.* LEFT: *An engraved nineteenth-century portrait of Johannes Kepler.*

As Kepler thought about the idea, talked with von Wackenfels, placated Emperor Rudolf, and began to sketch his story, inserting puns and allusions that von Wackenfels and Kepler's other friends would enjoy, he was well aware that this effort could serve as something other than a summer's pastime or an intellectual exercise. It would be, in a way, a Trojan horse—taking Copernican ideas, disguised as an imaginative myth that no one need take seriously, into the stronghold of those who still resisted Copernican astronomy. Its fancy would not be mere fancy, for this "dream" would be strongly underpinned by solid scholarship. Kepler's little book was becoming, for him, much more than a whimsical diversion. It was well in the mainstream of his lifelong effort to convince his peers that Copernicus had been right.

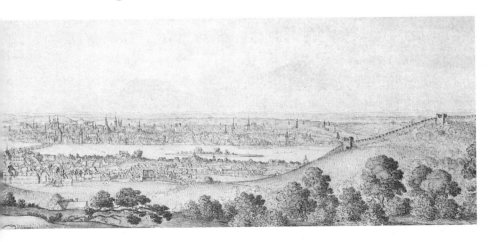

In the spring of 1610, a copy of Galileo's new book *Sidereus Nuncius* (or *The Starry Messenger*), which reported on Galileo's first observations with his new telescope, was sent as a gift to Emperor Rudolf, who lent the book to Kepler. Kepler's own copy arrived from Galileo a few days later, and Kepler was able to utilize information about the surface of the moon that Galileo had observed. The first mention of Kepler's project that appears in surviving documents is a letter he wrote to Galileo responding to *Sidereus Nuncius*. He did not win the friendship of the arrogant Galileo by pointing out in this letter that Galileo wasn't the first to deduce that there are mountains on the moon. Plutarch, fifteen centuries earlier, had upstaged him, and Kepler himself had used a technique from the ancient Greeks to estimate the elevations of the moon's mountains when he was a student at Tübingen.

Unlike Lucian of Samosata or Plutarch, Kepler had an excellent understanding of the reasons why a flight to the moon was impossible for anyone in his time; but as a theoretical exercise, he also clearly took great pleasure in working out the scientific problems of such a journey and how they might be solved. No whirlwinds or secret passages for him. Anyone who has undertaken—no matter how fleetingly—to write respectable science fiction will recognize how much more difficult such an exercise is than simply relating science fact. In a letter written twenty-four years after he and von Wackenfels first discussed the book, Kepler mentioned to a friend,

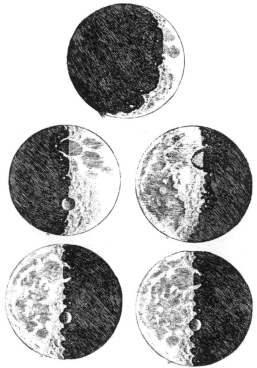

OPPOSITE: *Title page of*
Sidereus Nuncius (*The Starry*
Messenger) *by Galileo, 1610.*
LEFT: *Galileo's sketches of*
the surfaces of the moon from
Sidereus Nuncius.

Matthias Bernegger, that "In my [book], there are as many problems as there are lines, which must be solved partly by means of astronomy, partly by physics, partly by history." Since few people, he suspected, would think it amusing to solve them, he had by that time decided to solve all of them himself in extensive "notes."

The hero of *Somnium* is Duracotus, a name Kepler borrowed from Plutarch. In the story, Kepler dreams of reading a book he has discovered in the market. It is Duracotus's own account of his adventures. "My name is Duracotus. My home is Iceland, which the ancients called Thule." So it begins. Duracotus is feeling liberated to be able at last to write about an adventure that his mother, Fiolxhilde, forbade him to reveal while she was still alive. In a flashback in this story within a story within a dream, the curious boy, then age fourteen, accidentally destroys a little pouch of herbs and patches of embroidered cloth—presumably some sort of charm—that his mother has made and sold to a ship's captain. No longer able to produce the herbs and cloth as promised and unwilling to refund the captain's money, Fiolxhilde gives him Duracotus instead.

The ship sails the next day for Bergen, Norway, but is blown off course, and the captain decides to take advantage of the ill wind to deliver a letter to the isle of Hven (now called Ven) in the Øresund (the strait between modern-day Denmark and Sweden). The letter is addressed to the great astronomer Tycho Brahe, lord of the island. Kepler gives Tycho Brahe very good press in *Somnium*, though his real-life relationship with the arrogant genius had not been entirely happy by any means and took place not on Hven but in Prague. Duracotus learns to speak Danish with the help of Tycho's students, who come from all over the world, and he also shares wondrous whole nights of gazing at the skies with Tycho's fabulous instruments. This part of Kepler's story sounds itself like science fiction, but in truth it is an accurate description of what life at Tycho Brahe's extraordinary castle-observatory on Hven was like. Kepler never went there himself, but must have heard many stories from those who did.

After several years, Duracotus decides to return home to Iceland,

A seventeenth-century engraving of astronomer Tycho Brahe.

where he finds his mother still alive and eager to hear about his adventures abroad. Duracotus, equally curious about his mother's history and where she learned her "arts," finds that Fiolxhilde is well acquainted with several "very wise spirits." These are in fact *daimons*—a word of Greek derivation that can mean "one who knows" but has the alternate meaning of "evil spirit." Fiolxhilde tells her son that one particularly friendly daimon can be called forth by an incantation of twenty-one characters. As any reader can see (though apparently Kepler did not) a more-than-faint aura of witchcraft had entered the story. Kepler was being shockingly incautious for a man of his time. When he decided later to add helpful explanatory notes to the book, Note 38 would enlighten his audience: twenty-one is the number of letters in the words *Astronomia Copernicana* (Copernican astronomy) and the number of possible conjunctions between pairs of planets. Clearly he had been expecting very erudite, classically educated readers who loved puzzles and allegory.

The daimon invites both Duracotus's mother and Duracotus to accompany him to a place called Levania. As Kepler would explain in a note, "Levania" is a name derived from the Hebrew name for the moon: *lebana*. "Hebrew words, which are less often heard by us, are recommended for occult arts by the greater aura of superstition attached to them," he wrote. The fact that Kepler was making fun of those who practiced the occult was a subtlety not to be caught by many of his contemporaries.

Fiolxhilde tells her son that the journey to Levania, even if undertaken at the right time, season, and position of the moon, earth, and planets, is extremely dangerous for humans, though not for daimons. Very few humans are acceptable, but one category of human that is especially well-suited are "dried-up old crones." The joke here, Kepler's notes would say, was on his old mentor at Tübingen, Michael Mästlin, though surely this also meant that Fiolxhilde was well-qualified. The full line in Kepler's story reads "dried-up old crones . . . who since childhood have ridden over great stretches of the earth at night in tattered cloaks on goats or pitchforks." What was he thinking? In one of his notes, Kepler would write that this line, along with the conjuring of spirits in the story, was primarily to blame for his mother's witch trial.

From the point of Duracotus's and Fiolxhilde's departure for the moon, or Levania, the narrative is less a story and more a vehicle for conveying Kepler's intricate knowledge and understanding of astronomy and mathematics—and the correctness of Copernican astronomy. Everything that happens and that Duracotus observes is grounded in that understanding, and what is simply stated in the text is explained at length in the notes. The timing of the journey, though it might at first sound more astrological than scientific, is soundly based. Kepler realized that because both bodies are in motion, the shortest route to the moon from Earth would not be a straight line between the two but a line from the earth to a location in space where the moon and the voyagers would arrive at the same time. This, he knew, would be best achieved by having the entire journey take place in the short interval during which the moon is in the shadow of the earth during a lunar eclipse, the longest duration of which is four and a half hours. This way, the travelers could avoid deadly direct radiation from the sun. The daimon limits the journey in this shadow to four hours to be on the safe side.

The experience of weight and the necessary arrangement of a traveler's body to distribute the shock of takeoff; the length of days, years, seasons, and the weather and geography of the moon; the extremes (and lack of extremes) of heat and cold; the view of Earth (here called Volva) and the observed movements of the planets and stars from the moon; the dramatic differences between conditions on the opposite sides of the moon; the size and nature of the moon dwellers—all are described in a virtuoso display of Kepler's scientific prowess. Duracotus doesn't have any adventures on the moon or direct encounters with its inhabitants, as we would expect in modern science fiction. He is an observer only. But Kepler has set the standard for all the best science fiction in the centuries to come: the story may be wildly imaginative, but it must be underpinned by arguably plausible science.

Finally, in something of an anticlimax, Kepler awakens from his dream to find his head covered by a pillow and his body wrapped in the bedclothes, much as Duracotus and Fiolxhilde had to wrap themselves for their journey.

Kepler had no intention of publishing this story immediately. He wanted to include his own translations of Lucian of Samosata's *True Story* and Plutarch's *On the Face in the Moon*. Unfortunately, the year after Kepler's letter to Galileo, the first copy of *Somnium* was circulated privately. Kepler didn't know how this came about, but in his own account, introduced with the words "If I am not mistaken," a copy somehow reached Leipzig and from there was carried to Tübingen by one "Baron von Volckersdorff and his tutors in morals and studies." However it may have occurred, in 1611 *Somnium* was out there, public, no longer under Kepler's control, ending up in the hands of people who were ill equipped to understand it and who were likely to use it as a weapon against him. Kepler worried that the book and his family were becoming the subject of gossip: Was not Johannes Kepler himself Duracotus, prepared to take questionable and unseemly steps, even consult with demons, in the interest of his science? Wasn't the fact that the hero spends some time with Tycho Brahe certainly a clue that Kepler was writing about himself? Was not Kepler's mother herself Fiolxhilde? It was

here that the *Somnium* story veered dangerously close to autobiographical. Kepler had been exceedingly careless.

By Kepler's own description, his mother Katharina was a small, thin,

A c. eighteenth-century French school painting entitled A Horned Witch; *some believe it depicts Kepler's mother, Katharina, who was accused of being a witch.*

garrulous, quarrelsome, unpleasant woman with a dark complexion. She had a restive, inquiring mind much like her son's, which she mostly devoted to growing herbs and creating homemade mixtures for healing. What in the young Johannes had developed into a rich intellectual curiosity, in her, without the benefit of any education, had taken the form of obnoxious nosiness. Her neighbors and acquaintances called her an evil-tongued shrew. An aunt who had cared for her as a child had been burned at the stake as a witch. When Kepler based his character Fiolxhilde on his mother, his description erred on the positive side. But he almost condemned her to death.

Kepler was also incautious in assuming that Ptolemaic astronomy had by and large been overthrown in favor of Copernican astronomy. He voiced this assumption allegorically, and later admitted his error in Note #2 of *Somnium*, which footnotes the passage about Duracotus being free to tell his tale after his mother's death. Fiolxhilde, Kepler explains, represents Ignorance, but Science (Duracotus) is her offspring. As long as Ignorance lives, it is unsafe for the offspring, Science, to reveal the hidden causes of things. "Age must be respected, a ripening of years must be awaited, worn out by which, as if by old age, Ignorance will finally die." But "the ancient mother is still alive in the Universities." That was certainly not being diplomatic, and the old ideas were more tenacious than he thought.

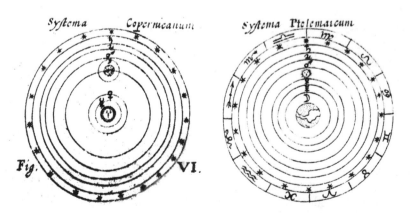

Copernican (left) and Ptolemaic (right) models of the solar system, from De revolutionibus orbium coelestium (On the Revolutions of the Heavenly Spheres, or Celestial Orbs) *by Nicolas Copernicus, which was published shortly after his death in 1543.*

Though Kepler was not happy in 1611 about his book circulating out of his control, there were no immediate repercussions, and he had other things on his mind. That winter in Prague, Kepler's three children contracted small-pox. Six-year-old Friedrich, who had been a particular delight to his father, died. During that same period, the streets around the Kepler house became a bloody battlefield when troops brought in by Emperor Rudolf clashed with local Bohemian–Protestant vigilantes. The emperor himself was showing more and more signs of madness. Kepler's happier days in Prague were at an end. He cut his losses by accepting a position as provincial mathematician in Linz. His wife Barbara died in June before they could make the move.

In Linz, Kepler wasted little time finding a second wife to care for his two orphaned children. The marriage and the subsequent birth of a

Belief in the dangers of witchcraft was still prevalent in seventeenth-century Europe, as evidenced by this fantastic etching entitled Zaubereÿ *(Sorcery). It was created in 1626 by the Swiss-born engraver Matthäus Merian the Elder after a drawing by Michael Herr.*

daughter were a brief happy interval, and Kepler began serious work again on his science. Then, four years after *Somnium* had gone out of his control, Kepler learned that news of its contents had indeed reached the Duchy of Württemberg, where he had spent his childhood, and that people there were speculating about its connection to his own personal story and that of his mother. In December 1615, news arrived from his sister Margarete that the elderly Katharina had been accused of witchcraft. On January 2, 1616, Kepler wrote to the officials in Württemberg requesting the documentation supporting the accusation. In the reply he learned that he himself was mentioned in the documents as being associated with "forbidden arts."

An old, skinny woman as meddlesome as Katharina who dabbled in herbs and folk medicine was a dead-on target for suspicions of witchcraft in seventeenth-century Germany. It required little more than grudges and gossip of neighbors for such a woman to end up facing a witch trial. The people with whom Katharina associated—rather, the sort who were *willing* to associate with her—were the dregs of society. The most malicious of these was Ursula Reinbold, a friend of Katharina who was a prostitute and frequently sought abortions, mostly at the hands of her barber surgeon brother but at least once with the help of one of Katharina's herbal potions. The friendship soured when Kepler's brother Christoph had a minor business dispute with Frau Reinbold. She was ill at the time—the result of a botched abortion—and chose to believe that the potion Katharina had given her three and a half years earlier had been a "witch's drink" and the cause of her present problems. She and her brother demanded that Katharina produce a "witch's antidote." Katharina, though completely lacking in social graces and any sort of everyday wisdom, was not lacking in intelligence. To acquiesce would have been to admit to practicing witchcraft. Even with Ursula's brother's sword at her throat, she refused, and she countered the witchcraft accusations by bringing a libel suit against Ursula Reinbold.

Kepler took on his mother's legal defense himself, well aware that his book *Somnium* could prove a deadly liability. A thin old woman, well-versed in folk magic, with the power to summon demons; a son who took advantage of this magic to visit the moon. . . . Kepler did not know what copies of his manuscript were now in circulation, and he agonized over whether one might have reached Leonberg, his mother's town.

The libel trial was postponed for a short time because the bailiff, Lutherus Einhorn, didn't want it revealed that he had been present when the sword was held to Katharina's throat. Then, when the proceedings were about to begin, a fresh incident occurred: While walking along a narrow path, Katharina had brushed against the clothing of a girl whose mother owed Katharina money. The girl claimed that since that touch, the pain in her arm had increased until she could no longer feel or move her hand. Einhorn brought in a medical consultant to examine the arm—none other than Ursula Reinbold's brother. "It is a witch's grip," he declared, "it has even got the right impression."

At this point Katharina tried to bribe Einhorn—a disastrous mistake, for he recognized an opportunity to get himself off the hook by suspending the libel case. He sent charges of "witch's drink," "witch's grip," and attempted bribery to the high council in Stuttgart. Kepler, his sister Margarete, and Margarete's husband, who was a village pastor, hastily bundled Katharina off to Margarete's house and from there to Linz. There Katharina was safe, though the council back in Stuttgart issued an immediate order for her arrest and "strenuous examination." A witch trial had begun.

Title page of Witches apprehended, Examined and Executed for Notable Villanies . . . , *1613, illustrating what could befall an individual accused of witchcraft.*

By late 1617, it seemed safe to take Katharina (she was not a congenial guest) back to her home in Leonberg. Double tragedy had struck Kepler's immediate family again—the deaths of his two-year-old daughter and of his dearly loved twenty-seven-year-old stepdaughter by his first marriage. A new baby died the following February. A grief-stricken Kepler turned his attention to a book he was developing that would eventually be *Harmonices Mundi*, or *Harmony of the World*. He had begun it many years ago while mourning the death of his first daughter. In the research and writing of *Harmony* he discovered his third law of planetary motion. His most significance contributions to posterity were now complete.

Katharina Kepler's problems had, however, only just begun. Thanks to a heads-up in 1618 from an old classmate on the University of Tübingen's law faculty—probably Kepler's old university friend, Christopher Besoldus—Kepler was forewarned when in the autumn of 1619 a counter-civil suit was filed against his mother for damages for poisoning Ursula Reinbold with the "witch's drink." This time Katharina's enemies had amassed no fewer than forty-nine counts against her: unnatural eerie behavior, riding a calf to death, uttering fatal "blessings" over infant children, causing pain without touching, the unexplained death of animals, trying to entice a young girl to become a witch . . . and asking a gravedigger for her father's skull so that she could have it set in silver for her son Johannes, the imperial mathematician. Unfortunately, the last accusation was true. Katharina had heard in a sermon about an archaic practice of creating goblets from the skulls of dead relatives, and that had sounded like a good idea to her. *Somnium* now hovered like a scythe over both Kepler's and his mother's heads. The case dragged on. Einhorn was still bailiff. In August of 1620, he and the Reinbolds succeeded in having the civil complaint changed to a criminal case.

Katharina, now seventy-four years old, was dragged out of her daughter's house and imprisoned in chains, with two guards. For much of her thirteen-month imprisonment, she was required to pay for the guards and the costs of her food and upkeep. The Reinbolds lodged a complaint that so much of her money was being used in this manner that there would be little

Witchcraft trials often employed various "tests" to prove a person was a witch. This Dutch engraving depicts an accused woman being weighed before her trial. Witches were believed to be lightweight, enabling them to fly; if the scale (often rigged) tipped in favor of the weights, it "proved" the person was a witch.

left for them when the trial was over. Kepler's brother Christoph arranged to have the trial, with all the expected spectacle and scandal, transferred to another district, but he was inclined to abandon the situation himself and try to save his own reputation. Margarete, somewhat against her own husband's advice, wrote again to Kepler in Linz, and Kepler appealed to the Duke of Württemberg for a delay until he could arrive and defend his mother himself.

In September 1620, with danger and disgrace hanging over them, he, his pregnant wife Susanna, and their eight-month-old son slipped away in the dead of night from their home in Linz and embarked on the grim, secret journey to Württemberg. Kepler didn't even tell his assistant, Gringalletus, where they were going or why. Better that the mystified people of Linz should think their mathematician had fled for good rather than have them know the shocking truth.

The trial dragged on for months with more and more lawyers, wit-nesses, and written arguments, for Kepler insisted that all the defense law-yer's arguments be put in writing, even though that meant great expense in what was probably already a lost cause. Kepler and his own lawyer in Stuttgart put together a 126-page legal brief, much of it in Kepler's own handwriting, that addressed the charges one by one. When the trial ended in August 1621, all the proceedings had to be sent to the law faculty of the University of Tübingen, whose duty it was to decide the case. Even the force of Kepler's presence throughout the trial, his reputation, his skill in devising the defense, and a powerful friend on the faculty did not save Katharina. The court declared that she be shown the instruments of torture in the hope that her response would settle the case. It did, in a way that surprised Kepler as much anyone else. When the torturer displayed his instruments, described how he would use them, and commanded her to tell the truth, Katharina pulled together her aging wits, summoned the eloquence and religious cer-tainty she had bequeathed to her son, and saved herself. The report reads:

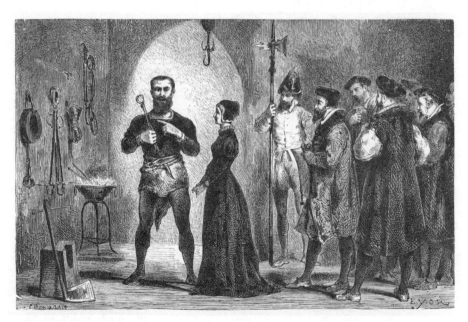

Undated engraving of Katharina Kepler being shown instruments of torture by the executioner.

She announced one should do with her what one would. Should
one pull one vein after another out of her body, she knew that she
had nothing to say. She fell to her knees, uttered the Lord's Prayer,
and declared that God should make a sign if she were a witch or
a demon or ever had anything to do with sorcery. Should she be
killed, God would see that the truth came to light and reveal after
her death that injustice and violence had been done to her, for she
knew that He would not take His Holy Spirit from her but would
stand by her.

The charges were dismissed, the Reinbolds were fined ten florins, and Katharina was set free; but by then she was a broken woman, a remnant of her former, cantankerous self. She died the following April. His mother's lengthy, demeaning, life-sapping ordeal was something that Kepler would not be able to forget, nor could he dismiss the suspicion that he was perhaps partly responsible for it.

As soon as Katharina was acquitted, Kepler hastened back to Linz, where his wife had already returned with their two young children. Susanna had given birth to a daughter. But for the Keplers, life in Linz was not to be peaceful. In Prague, Kepler's patron Emperor Rudolf II had been succeeded briefly by the Emperor Matthias and then by Ferdinand II, the very man responsible for the Counter-Reformation edicts that had forced Kepler out of Graz many years earlier. Ferdinand's treatment of Protestant leaders was brutal and grisly, though less so in Linz, where new statutes dealing with Protestants were more annoying and demeaning than deadly. Ferdinand even reconfirmed Kepler's appointment as imperial mathematician. In spite of these interruptions and the death of another child—his young son Sebald—Kepler was able to continue laboring on the *Rudolphine Tables*. This undertaking, begun before the death of Tycho Brahe and based on Tycho's observations, was also now benefitting from Kepler's own much more sophisticated and correct understanding of the solar system. When completed, the *Tables* would allow a user to calculate the position of any planet at any time thousands of years in the future or the past.

Kepler also occasionally turned his attention to *Somnium*. Early in December 1623, he wrote the letter to his friend Matthias Bernegger in which he spoke of his decision to add extensive notes to the book.

> *Two years ago, immediately after my return to Linz, I started to work again on the astronomy of the moon, or rather to elucidate it by remarks. . . . There are just as many problems as lines in my writing, which can only be solved astronomically, physically, or historically. But what can one do about this? People wish that this kind of fun, as they say, would throw itself around their necks with cozy arms; in playing they do not wish to wrinkle their foreheads. Therefore, I decided to solve the problem myself, in notes, ordered and numbered.*

Completing the *Rudolphine Tables* had to come first, and he finished them in 1624.* The ordeal of getting the *Tables* printed was the next difficulty. Astronomical type adequate for the *Tables* had to be custom-made. When the Thirty Years' War reached Linz, Kepler was obliged to open his house, which was located on the city wall, for soldiers guarding the wall. During the two-month siege, a fire consumed the printing press and only barely spared the handwritten manuscript of Kepler's *Tables*.

When the siege was lifted, Kepler, now with two more young children and his five-year-old daughter, sought permission from the emperor to leave Linz. By boat and wagon "laden with plates of my figures and Table work," he, Susanna, and their young family traveled up the frozen Danube to Ulm. There, under Kepler's careful supervision, the *Tables* were finally printed. In 1528, under the auspices of Albrecht Wallenstein, ruler of the Silesian Duchy of Sagan (also called Żagań), the Keplers moved to Sagan, where they would be allowed to maintain their Protestant faith and where Kepler could also remain in the service of the emperor. Kepler was now producing ephemerides,

---

* In addition to making it possible to calculate any planet's position for any time thousands of years into the future or the past, Kepler included logarithm tables—Tycho Brahe's catalog of his observations of a thousand stars along with the latitudes and longitudes of many cities.

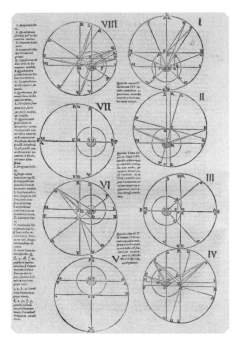

Kepler's illustrated discussion of the computation of planetary orbits in his Rudolphine Tables, *published in 1627.*

which provided a good part of the Kepler livelihood.* But shortly after the birth of his youngest child, Anna Maria, when Kepler was unable to leave his wife's side long enough to oversee the printing press, he directed his printer to work instead on printing *Somnium*. The printing continued only sporadically after Kepler was back in supervision. However, when Bernegger asked in the spring of 1629 whether he could recommend a mathematics textbook appropriate for use in a classroom, Kepler first listed a few choices and then wrote:

> *But what if I added, as a jest, my 'Astronomy of the Moon or of the Celestial Phenomena as Seen from the Moon?' This will provide the fare for us [Protestants in the Catholic Counter-Reformation] who are being chased off the earth as we travel or migrate to the moon. To that little book I am appending Plutarch's* Face of the Moon, *newly translated from the Greek by me.*

---

* These were tables, produced yearly, that gave the position of each planet for every day of the year. They were much in demand for both navigators and astrologers.

This final *Somnium* resembled a combination of *Star Trek* and *The Physics of Star Trek* by Lawrence M. Krauss, for there were fifty pages of notes and diagrams accompanying the twenty-eight-page story. Kepler felt safe now, long after his mother's witch trial and death, in pointing out in the notes how the book might have been used out of context to condemn her as well as embroil himself. But the notes contain much more and clinch the argument that this was not merely fiction but *science* fiction. For example, they show that Kepler understood the concept of gravity better than he is often given credit for based on his other works. He described clearly the point between the earth and moon where their separate gravitational attractions exactly cancel out.

Between printing runs of ephemerides, Kepler continued to print *Somnium*, but it was not finished when he left on a lengthy journey in early October 1630. He traveled first to Leipzig, carting his freshly printed books to sell at the autumn book fair, then to Linz in an effort to collect debts owed him there, and finally to a meeting in Regensburg, where the fate of the friend who had brought him to Sagan, Albrecht Wallenstein, hung in the balance. With Wallenstein's fall from grace with the emperor, the future for Kepler again looked bleakly uncertain. When Kepler rode into Regensburg on a useless old horse, he was exhausted and ill. He died there on November 15, 1630.

Kepler's son-in-law, Jakob Bartsch, continued the printing of *Somnium*, though it was still not quite complete at the time of Bartsch's own death. Kepler's severely impoverished widow Susanna was still trying to sell incomplete copies of it and sought the assistance of Kepler's son, her stepson, Ludwig when she moved to Frankfurt am Main in 1634. Ludwig, as a student, was in no position to offer her much help. He was still a candidate for a doctorate in medicine.

> *For two years I had heard no news about the condition of my relatives. From Frankfurt I wrote to them in Lusatia to let me know whether they were still alive and how they were faring. Lo*

and behold, my stepmother [Susanna Kepler], an impoverished widow with four orphans, comes to me at a turbulent time and in a place most inconvenient on account of the high cost of living. She brings with her the unfinished copies of this Dream, *and asks for my help. . . . In particular she wants me to complete the copies of this* Dream. *. . . I could not refuse this request, but instead I have made it my goal.*

Somnium, finally with a title page and dedication, was published in Latin as a complete book in 1634 in Frankfurt. The book never received any attention from the select few for whom Kepler originally intended it. It failed to earn the Kepler family very much money, though it was published the year after Galileo's famous trial, when there was a surge of public interest in Copernican sun-centered astronomy.

Today I am aware of only four English translations, one unpublished and the other three out of print. Science historians and modern–science fiction writers have largely forgotten *Somnium*. Might it have been lost entirely had it not contributed to the ordeal of Katharina Kepler? Perhaps,

but Jules Verne knew of it in the nineteenth century and so did H. G. Wells a little later, and both carried on superbly the tradition that Kepler had begun.

This statue of Katharina Kepler, as envisioned by sculptor Jakob Wilhelm Fehrle, 1937, stands at a village well in Leonberg, Germany.

ALAME

Typ. Plou frères.

NINE

Milutin Milanković

ASTRONOMY ON ICE

(1864–present)

ILUTIN MILANKOVIĆ WAS A MAN WITH A passion for adventuring in "distant worlds and times," not by ship or sledge but on the wings of mathematics and science. He also was a man who didn't mind being unrecognized and forgotten. So confident was he that his conclusions were correct that he confessed to welcoming lack of attention during his lifetime and even long after his death but fully expected, eventually, to receive "compensation from posterity—with interest!"

Today, nearly sixty years after his death in 1958, what has come to be known as the Milankovitch Astronomical Theory, which links the waxing and waning of Earth's ice ages to changes in Earth's orbit around the sun, is part of the canon of geophysics. It was one of the most important advances in the twentieth century. Yet many, even in Milanković's own field, know nothing about the man. Most modern geophysicists would have to look up his first name, his nationality, and when and where he lived. They refer to the "Astronomical Theory" usually without a name attached. Outside the field, most scientists shrug if you mention him. He is unknown.

OPPOSITE: *The Matterhorn, from an illustration by Swiss artist Rodolphe Töpffer, 1854.* LEFT: *Portrait of Milutin Milanković by Serbian painter Paja Jovanović, c. 1945.*

I first encountered his story in Cambridge, England, in a college lunch conversation with Nick Shackleton, eminent geologist and paleoclimatologist and great-nephew of the famous polar explorer Ernest Shackleton. Nick shares the credit for pulling Milanković's theory out of the "rejected and forgotten" drawer in the late 1970s. He urged me to write Milanković's biography. "He did some of his most important work as a prisoner of war during the First World War, you know." Indeed he did, when his honeymoon took him and his bride to the wrong place at the wrong time. But Milanković was surely one of the most imperturbable, serenely focused of men. He described his first day in prison:

> *The heavy iron door closed behind me. The massive rusty lock gave a rumbling moan when the key was turned. . . . I adjusted to my new situation by switching off my brain and staring apathetically into the air. After a while I happened to glance at my suitcase. . . . My brain began to function again. I jumped up, and opened the suitcase. . . . In it I had stored the papers on my cosmic problem. . . . I leafed through the writings . . . pulled my faithful fountain pen out of my pocket, and began to write and calculate. . . . As I looked around my room after midnight, I needed some time before I realized where I was. The little room seemed like the nightquarters on my voyage through the universe.*

To understand Milanković's voyage, we must look back to the mid-nineteenth century and another forgotten man. In 1864, forty-three-year-old James Croll was a janitor at the Andersonian College and Museum in Glasgow (now the University of Strathclyde). Janitorial work left his mind free to wander. He began ruminating about the ice ages he had read about and decided to try to find out what had caused them. The idea was not promising. Before becoming a janitor, Croll had worked unsuccessfully as a mechanic, carpenter, tea shop proprietor, salesman, hotel keeper, and insurance broker. There is no evidence that he was any more successful as a janitor. His talents, to say the least, lay elsewhere, in philosophy, theology, and physical science. He was totally self-taught . . . and well taught.

Photograph of Louis Agassiz at the blackboard, c. 1870.

The idea that the earth had endured ice ages in the past was relatively new. As recently as 1837, Louis Agassiz had presented a controversial theory of a Great Ice Age at a meeting of the Swiss Society of Natural Sciences. Agassiz had been puzzling over the scratched, many-faced, jutting rocks in the surrounding mountains; boulders lying far from their areas of origin; and other subtler features such as the rounded shapes of pebbles. He was convinced that an ancient age of ice had been responsible. He speculated that Earth had been "glaciated" at least from the North Pole to the Mediterranean and perhaps even entirely covered with ice. Five years after Agassiz's presentation, ideas and calculations were falling into place that would make the success of Croll's plan not completely unlikely. Frenchman Joseph Adhémar was theorizing that there was a link between the way the earth moves around the sun and ice ages on Earth, and French astronomer Urbain Le Verrier showed that the eccentricity of Earth's orbit (how elongated its elliptical orbit is) is slowly but continuously changing due to the gravitational pull of the other planets in the solar system. Le Verrier had calculated the history of Earth's orbit over the past hundred thousand years.

239

Statue of Urbain Le Verrier at the Observatoire de Paris by nineteenth-century French sculptor Henri Chapu.

Croll set himself the prodigious task of using formulas developed by Le Verrier to carry that calculation back to three million years ago. He illustrated the changes on a graph, showing periods lasting tens of thousands of years when Earth's orbit's eccentricity was at its most elongated, alternating with intervals of low eccentricity when Earth's orbit was almost a circle (see Figure 1). These changes were approximately in sync with what geologists of Croll's time believed to have been the advance and retreat of ice sheets on Earth. It seemed there must be a link.

There was a problem with this line of thinking. Le Verrier had found that changes in orbital eccentricity do not much affect the total amount of heat the earth receives from the sun over the course of a full year. How, then, could those changes cause ice ages? Croll thought he had the answer: Regardless of the *total* amount of heat reaching the earth from the sun annually, less sunlight in winter means greater accumulation of snow. Snow reflects sunlight back into space. When the area covered by snow increases even a little, more sunlight is reflected back. The focus of inquiry, he thought, should be on astronomical factors affecting the amount of sunlight Earth receives in winter, when snow accumulates. With this, a new player entered the field: the *precession* of the equinoxes.

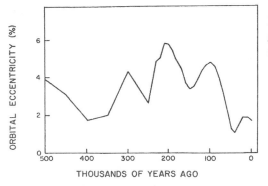

Figure 1: *James Croll's graph illustrating changes in orbital eccentricity.*

Viewed in the night sky from the northern hemisphere, the stars appear to rotate around a point near Polaris. We call Polaris the polestar because the earth's axis of rotation (a line drawn from pole to pole) points toward it—or, easier to visualize, the North Pole points toward it. In 2000 BCE, Polaris was not the polestar. At that time, all the stars appeared to rotate around a point between the Little Dipper and the Big Dipper. Two thousand years earlier, in 4000 BCE, the stars appeared to rotate around the *tip of the handle* of the Big Dipper. The reason for these changes is that the earth wobbles like a spinning top that is slowing down, and, like the knob on top of such a toy, the North Pole draws a circle in space (see Figure 2). Because Earth bulges

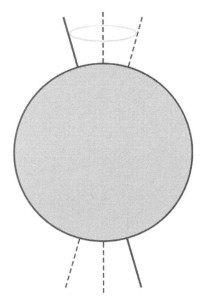

around the equator, there is extra gravitational pull of the sun and the moon on that bulge, and this extra pull, along with subtler factors, causes the wobble. The wobble is extremely slow. Twenty-six thousand years must pass before the earth's axis of rotation gets back to the same point on the circle.

Figure 2: *The wobble of Earth's axis of rotation.*

Owing to this wobble and other astronomical movements, the positions where Earth is in its orbit at the equinoxes (when the sun is directly overhead at the equator, the two poles are equidistant from the sun, and day and night are of equal length all over the earth) and at the winter and summer solstices shift around the orbit. Today, though it might at first glance seem counterintuitive, the northern hemisphere has winter when the earth is closest to the sun.

Clearly the precession of the equinoxes could change the intensity of the seasons, but that was not the whole story. Croll saw that variations in the eccentricity of Earth's orbit affect *how much* the precession changes the intensity of the seasons. If Earth's orbit were circular (with zero eccentricity), the earth would be the same distance from the sun no matter where in its orbit the winter solstice occurred. The precession wouldn't affect the climate at all. Winters would never be cold enough for an ice age. That is in fact the way we experience the climate today. But when the elliptical orbit is very greatly elongated, there will be exceptionally warm winters if the winter solstice occurs near the sun and exceptionally cold winters if it occurs far from the sun (see Figure 3).

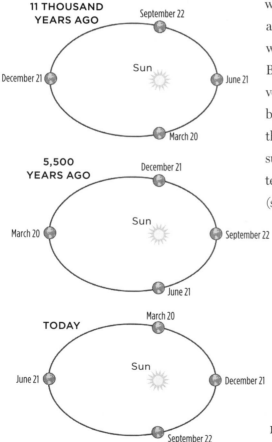

Figure 3: *Precession of the equinoxes.*

Croll concluded that an ice age requires two things to happen simultaneously: (1) winter solstices occurring at an orbital point farther from the sun; and (2) high orbital eccentricity. Croll saw these factors working together. Neither, by itself, would suffice. When Earth is sufficiently far from the sun on December 21, an ice age can occur in the northern hemisphere. When Earth is sufficiently far from the sun on June 21, an ice age can occur in the southern hemisphere.

Opportunities for ice ages alternate between the two hemispheres in a cycle that takes twenty-two thousand years to complete. But only during certain long intervals when the orbital eccentricity is great enough do these potential ice ages actually occur. These intervals are *glacial epochs*. In between them are *interglacial epochs*. According to this way of thinking, the most recent glacial epoch must have run anywhere between 250,000 and 80,000 years ago, and we are now in an interglacial epoch, when Earth's orbital eccentricity is too low for an ice age to occur no matter where the winter and summer solstices occur on the orbit.

But could these astronomical changes, taking into consideration the reflection of sunlight off accumulating snow, actually account for ice sheets covering most of Europe and North America? Croll was troubled. Perhaps

they served as a triggering mechanism to set off responses within Earth's own climate system, such as shifts in trade winds and the flow of ocean currents. Croll hypothesized that the tilt of Earth's axis might also play a role.

Engraved portrait of Scottish scientist James Croll, 1897, from Popular Science Monthly.

Croll died in 1890 at the age of sixty-nine. His theory of the ice ages had at first made a profound impression on the scientific world but fell into disfavor with the discovery that the most recent ice age ended a mere ten thousand years ago, not eighty thousand, and calculations by meteorologists showed that the mechanism Croll proposed could not have had a sufficient effect on the climate. By the turn of the century, his work had become a historical curiosity, on its way to oblivion.

In 1896, seventeen-year-old Milutin Milanković said farewell to several doting generations of his family who had come out to watch his train leave the station in their village of Dalj in Croatia. He was heading to Vienna to begin a five-year course in civil engineering at the Vienna Technical High School. It took nearly a year for the small, frail teenager to adjust to the lifestyle of a great European city, and he enjoyed the unaccustomed freedom a little too much. Nevertheless, by the end of his second year Milutin

had buckled down to his studies with excellent results. He was also organizing his time so as to enjoy the cultural life of Vienna, learn about its architecture and art, haunt its museums and libraries, and attend the theater, concerts, and the opera in the evenings. Opera became his favorite—"his greatest love."

Milanković as a student, c. 1900.

Milutin would have completed the engineering course in the requisite five years had he not made an error in a final assignment that took two months to correct. It was characteristic of him that, instead of being frustrated, he was rather pleased and spent the extra months broadening his knowledge of his subject as he had not been able to do during the regular course schedule.

After graduation, Milutin served an obligatory twelve months in the military. Heading off to an officers' school in Zagreb, he hoped that "my present military service does not kill all my human initiative and independence and make out of me a robot." It did not. Soon after completing the year, he borrowed money from an uncle and returned to Vienna to pursue a doctorate in engineering. He focused on a new building material, reinforced concrete, and also on a different kind of "mechanics"—astronomy, or "celestial mechanics."

At age twenty-five, Milutin had his doctorate but no success from six applications he had submitted for engineering positions. Returning home to Dalj, he prepared to settle down there for the rest of his life, not dissatisfied with the prospect of spending some time pursuing his growing interest in astronomy. Then one of the job applications proved belatedly successful, and he was back in Vienna, solving some tricky engineering problems, obtaining several patents, and making a fine reputation as an engineer. Astronomy, however, was far from forgotten. In 1905, traveling regularly to the site of a large modern sewerage plant he was designing for the city of Belgrade, he noticed a placard listing courses being offered at Belgrade's New University. One course offering made him long to be a student again—applied mathematics, which included theoretical physics and celestial mechanics. Four years later, however, he would not be a student in that course. He would be teaching it.

The large engineering projects in which Milanković was meanwhile engaged required considerable theoretical knowledge and gave him good reason to spend time in the library at the Vienna Institute of Technology. His enthusiasm for science was growing exponentially. He added more advanced applied mathematics to his areas of expertise and also more astronomy and theoretical physics. When the University of Belgrade offered him a place on their Philosophical Faculty as Chair in Applied Mathematics in 1909, the thirty-year-old Milanković surprised his sophisticated Viennese friends by eagerly moving to "backwater" Belgrade.

Photochrom print of the seat of the University of Belgrade, c. 1900.

Almost entirely self-taught in much of the subject matter he was teaching, Milanković was pleased to find that in the first year he would only have to develop courses in "rational mechanics"—an area he knew well and had utilized in his engineering. He had a year to bring himself up to standard in celestial mechanics and still another to do the same in theoretical physics. After three years, he was finally certain he had made the right decision and was properly equipped for his professorial position. His formal education in Vienna had given him "solid knowledge from my excellent teachers," and his "work on science as a sideline to my engineering work" had prepared him well. "I felt I had not done badly."

Milanković had also learned a profound lesson about science. He wrote in his memoirs,

Successful mastery of such knowledge does not make one a scientist. It is the search for new, unsolved problems and the finding of their solutions that makes the real challenge. In order to be able to pose such questions one needs intuition and scientific ability. I thought I had both.

And so, secure in his new academic position (with a salary only one-seventh of what he had been earning as an engineer) and ever optimistic, he set out to search not for answers but for a major unsolved problem. He tried the trendy new area of Einstein's theory of relativity but found it too crowded with people all attacking the same questions. Although he would continue to work on it privately for the rest of his life, living in Belgrade meant he was out of the mainstream, and it would be impossible to keep ahead of that game. Then, while searching through works in the field of meteorology, he came across papers dealing with the spread of heat from the sun over the surface of the earth and discovered, one after another, that these papers and their conclusions were seriously flawed. In 1913, he published his own paper on the subject, and this paper was, as he put it, "to become the founding stone of my future work." He had found his problem, and he put the words in full caps in his memoirs: THE ICE AGES.

Milanković knew he was taking on a puzzle that was widely thought to be insolvable. He began methodically, by dividing it into two parts. The "astronomical part" would concern the way the movement of a planet in orbit creates and affects the cycles of day and night and seasons. He was particularly interested in the way the gravitational pull of the planets on one another changes their orbits and tilts over very long periods of time and, as a consequence, changes the length of seasons and periods of exposure to the sun on their surfaces. He was determined not merely to speculate about such activities and changes but to represent them by mathematical formulas. The "physical part" targeted the creation of a mathematical theory with which he could numerically trace the effect of the sun's rays on the atmosphere and the surface of the earth. He hoped that with both parts in place, he could discover the whole mechanism of Earth's climate and would be able to describe the climate in ancient times, perhaps solving the mystery of the ice ages. This was not to be a quick study. It involved problems spread among several sharply divided branches of science. Milanković was in his element.

In the winter of 1913/14, Milanković fell in love with Christine Topuzović, a young woman from an old merchant family in Belgrade. He described her as "intelligent and very beautiful." She was thirty years old when they married in the late spring of 1914. He was thirty-five. Their honeymoon began in his hometown of Dalj, so that she could meet his family. From there they planned to travel to Vienna and Switzerland. Two weeks after their wedding, when they were ending their stay in Dalj, those plans changed dramatically. On June 28, 1914, Archduke Franz Ferdinand, heir to the throne of Austria, and his wife Sophie were assassinated in Sarajevo by a Bosnian student educated in Serbia. A month later, Austria-Hungary declared war on Serbia. Milanković, a Serbian citizen, was taken as a prisoner of war.

Portrait of Archduke Franz Ferdinand with his wife Sophie, Duchess of Hohenberg, and their three children: Prince Ernst von Hohenberg; Princess Sophie; and Maximilian, Duke of Hohenburg. The assassination of the Archduke and his wife during a state visit to the Bosnian capital, Sarajevo, on June 28, 1914, precipitated the First World War.

Christine, or "Tinka" as he called her, was a brave, resourceful, strong-willed woman. When Milanković was transported to Karlovać, in Croatia, she followed and stayed with friends there so that she could visit him regularly. When he was moved to a large camp in Nezider, Hungary, she followed again. Meanwhile she also traveled widely, attempting to use all possible contacts to help him. When she approached his old professor of mathematics at the Institute of Technology in Vienna, Emanuel Czuber, she was able to persuade Czuber to act on her husband's behalf not by arguing on the basis of friendship and fondness for a former student but by pointing out that Milutin would not be able to work on science in the camps. To the old scientist, this was unthinkable. Czuber was unaware that his imperturbable pupil was enjoying mental trips through the universe from his prison cell and calculating and writing much as he would have at home.

Czuber was well-placed. His son-in-law was a Habsburg archduke with high-level connections in the Austrian military. On Christmas Eve, 1914, Milanković's nearly five-month incarceration came to an end. Given a choice of living under house arrest either in Vienna or Budapest, he chose Budapest, for in Vienna "everyone was starving." He was ordered to report to the police station once a week, all his correspondence was scrutinized, and travel was forbidden without special permission. An uncle from Dalj brought him his books and manuscripts and opened a bank account for him in Budapest. Milutin and Trinka would have to live on part of Milutin's inheritance, but he was sure he would eventually be able to "transfer my scientific knowledge into sound, lasting, intellectual capital." When their resources began to run out, he took a temporary job with a Swiss civil engineering consultancy, designing a hospital in Slovakia. Otherwise—having made his presence in Budapest known in the library and mathematics department of the Hungarian Academy of Sciences—he was given the assistance he needed to continue his work, and he did so largely undisturbed. Trinka gave birth to a son on Christmas Day, 1915. The three of them lived in Budapest until the end of the war and in March 1919 finally took the Danube steamer back to a sadly war-damaged Belgrade.

Milanković and his family were almost penniless, but the war had not much harmed the family home that Tinka had left in 1914 to get married, never suspecting that she would not return for five years. An offer came almost immediately from the Yugoslav Academy of Sciences and Arts to publish the work Milanković had completed in Budapest. Because of the current political situation, it would have to be translated into French. The book *Théorie mathématique des phénomènes thermiques produits par la radiation solaire* was published in 1920.

Milanković had not been wasting his time while in prison and under house arrest. He had succeeded in describing mathematically the present climates on Earth, Mars, and Venus and had demonstrated that changes in the geographic and seasonal distribution of sunlight, due to astronomical variations, could produce ice ages. His mathematical description of Earth's past climate solved the mystery of how and why it had changed. He had successfully tackled the question of which latitudes and which seasons are most critical to the advance and retreat of the ice sheets and which astronomical changes are the main causes of ice ages.

Milanković concluded, as Croll had speculated earlier, that Earth's orbit is subject to slow, centuries-long changes that affect how much of the sun's radiation reaches different parts of the globe. Three features of the orbit vary over time: (1) the position of the equinoxes (see Figures 2 and 3); (2) the eccentricity of Earth's orbit (Figure 1); and (3) the tilt of Earth's axis of rotation relative to the plane of its orbit (Figure 4).

We have already seen the meaning and effect of the precession of the equinoxes and the eccentricity of Earth's orbit, but the third element—the way changes in the tilt of Earth's axis of rotation in relation to the plane of its orbit affect the possibility of ice ages—requires further explanation. The tilt at present is 23.5 degrees. When the tilt is less than that, the poles receive less sunlight than they do today. When the tilt is more than that, the poles receive more sunlight. If Earth's axis of rotation were at right angles to the plane of Earth's orbit, the poles would not be exposed to the sun at any time of year. The potential for an ice age would be at its maximum. If it were tilted at fifty-four degrees, everywhere on Earth would have the same annual amount of sunlight. The amount of tilt runs in a forty-one-thousand year cycle and never reaches those extremes.

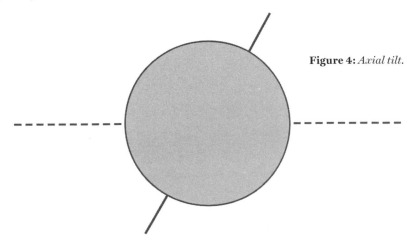

Figure 4: *Axial tilt.*

Milanković had at his disposal, as Croll had not, calculations by Ludwig Pilgrim of Germany of variations in the three astronomical properties over the past one million years, showing that the changes in eccentricity followed a 100,000-year cycle; changes in axial tilt a 41,000-year cycle; and the precession of the equinoxes a 24,000-year cycle. As we will see later, more recent calculations and studies of the long-term variations of these three astronomical parameters over the past show that their periods are not as rigidly constant in time as these numbers suggest, but they were the best data available to Milanković, and they served him well.

Understanding the elaborate choreography in which these three properties interact and affect the climate on Earth required the kinds of mathematics and engineering skills that were Milanković's forte. He confirmed what Croll had foreseen: that a decrease in axial tilt leads to a decrease in summer radiation, and a decrease in the earth-to-sun distance at any season means an increase in radiation during that season. But Milanković found subtler effects as well. At the poles, the influence of the axial tilt is large, but that influence becomes small toward the equator. Near the equator, the influence of the precession cycle is large, but that influence is small at the poles.

In his memoirs, Milanković described the scope of his book:

> *It describes and explains in precise mathematical terms how the
> Sun emits light and thermal rays into interplanetary space; how
> caloric units carried by these rays spread over the atmospheric
> mantle of the Earth and other planets; how the Sun's rays pene-
> trate these atmospheric mantles and reach the surface of the Earth
> and the planets; how the Sun's rays illuminate and heat these
> atmospheric mantles and planetary surfaces, and how these rays
> create daily and yearly temperature oscillations on the planetary
> surfaces and within the various atmospheric layers.*

Milanković was merciful to the readers of his memoirs: "I would not
be able to explain [for them] the mathematical means by which I solved
this cosmic problem. It will have to suffice that I describe just the end
result."

As conditions returned almost to normal after the war, life in Belgrade
was a good life. Milanković taught young people from a very wide range of
backgrounds who had fought in the war, and it was an interesting challenge
to prepare a curriculum that would meet the needs of those with so many
different levels of knowledge. He and his colleagues decided to reduce the
pre-war course to comprise what they considered essential facts, "a good
basic education being better than a poor broad one," he concluded. "The
difficult times through which this war generation of students had gone had
not only made them tough and mature but a class of excellent students."

It became the pleasant custom in Belgrade for families to hold recep-
tions once a month where friends would gather over tea and cakes and discuss current affairs—which Milanković revealed mostly meant "local gossip."
Milutin and Trinka did better than that. They held a reception every Sunday
in the combined study/living room of their home with an open invitation to
their old acquaintances and scientific and musical colleagues from the university, including some they had met earlier in Budapest, and new Russian
friends. Belgrade was full of Russian immigrants, many of whom were now

professors at the University—impoverished but highly respected for their science and their intellectual and personal values.

Milanković was at last, between the two world wars, living the ideal life of a European intellectual. These years were, in his own estimation, the most productive in his career. He did most of his work at home in his book-lined study, occasionally spending the day at the university. Twice a week he lectured there, then walked to his club to meet friends. Dinner at home was at eight o'clock sharp and lasted two hours, accompanied by discussion of world affairs and music. Afterward he read for an hour, then turned off the light and sat in the dark to think. Occasionally he took on an engineering project. These, he reported, tired him out physically but "did not slow down my scientific (and literary) work. Such work needs intervals in which one can recharge his batteries." He hadn't stopped thinking about ice ages and would continue to apply and refine his theory for many years.

In September 1922, a letter arrived from the German climatologist Wladimir Köppen. Köppen had read Milanković's book and pointed out that though Milanković's astronomical theory coincided successfully with what geologists knew about ice ages over the past 130,000 years, there

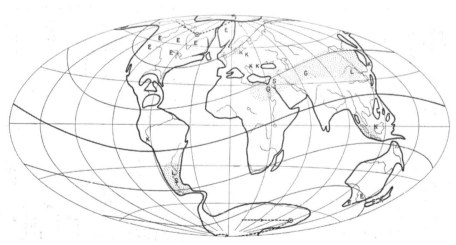

A climate map from German climatologist Wladimir Köppen. Köppen and Milanković worked together on a project to explore the climate of Earth's remote past.

had been ice ages before that, spread over at least 600,000 years. Would Milanković be willing to undertake calculations stretching that far back and collaborate with Köppen and his son-in-law, Alfred Wegener, on a large project to explore in depth the climate of Earth's remote past? Milanković enthusiastically agreed and swept everything off his desk except for his copy of his own book, which contained all the theory and information he would need for his part of the project. He already had the formula that allowed him to calculate for any past year, even 600,000 years ago, the amount of heat that any predetermined parallel on Earth received from the sun during the summer half of the year.

Milanković's part of the project was to concentrate on astronomy and calculate the changes in the amount of solar radiation reaching the top of Earth's atmosphere over the past 650,000 years. He was not to take into consideration the influence of Earth's atmosphere itself. He would focus his study on latitudes 55 degrees, 60 degrees, and 65 degrees north, paying particular attention to the amount of radiation in the summer half-year.

The choice of the *summer* half-year might seem unlikely, but Köppen and Milanković disagreed with Croll's conclusion that changes in winter radiation are the crucial factor in determining when an ice age occurs. It seemed obvious to them that such changes must not greatly affect the annual "snow budget,"* because even in modern times temperatures in Arctic regions are cold enough for snow to accumulate. However, also in modern times, glaciers melt in summer. It is the diminution of heat during the *warmer* half-year that is the decisive factor. When the intensity of summer sunlight decreases, there is a decrease in this melting, and the annual "snow budget" becomes positive, tipping in the direction of glacial expansion. When summers in succession are cooler, the snow line (the elevation above which there is snow all year long) descends from the tops of the mountains and glaciers and moves into the valleys. This additional snow cover reflects the sun's rays and warmth back into interplanetary space, accelerating the spread of the ice sheet.

---

* A snow budget is the difference between the amount of snow that falls during a given period (usually a year) and the amount lost to melting or evaporation—thus, the amount added to or subtracted from the snowpack.

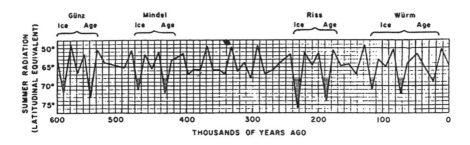

Figure 5: *Milanković's graph showing the radiation curve for latitude sixty-five degrees north.*

Milanković studied the altitude of the snow line at the different latitudes, and, keeping in mind that snow fields reflect incoming solar radiation, he formulated a mathematical relationship between summer radiation and the snow line altitudes. He was able to calculate how much increase in snow cover ought to result from any change in summer radiation from the sun.

Milanković produced a graph (see Figure 5) to illustrate how his theory predicted the way the intensity of summer sunlight must have varied over the past 600,000 years. It was a triumph (though no surprise to him) to find that the low points on the curve coincided with ice ages that experts knew had occurred in Europe. The names at the top of the graph—Günz, Mindel, Riss, and Würm—are the names that had been given to the four great ice ages.

Köppen summed up the triumph of this achievement a few years later:

> Other scientists—Adhémar, Croll, Ball, Ekholm and Spitaler— attempted to solve the problem of astronomical causes of climatic changes. None of them came up with acceptable results. Our greatest climatologist and my friend Hann, now deceased, discarded it out of hand. You [Milanković], not discouraged by his verdict, tackled the problem and solved it.

Milanković brushed aside this praise, insisting that it was Köppen who had solved it. "Yes, with your calculations!" replied Köppen.

After the 1924 publication of his book with Köppen and Wegener—
Die Klimate der Geologischen Vorzeit—Milanković's study of the amount
of the sun's rays reaching different areas of the earth (insolation) was
widely known and discussed. Work by German geologist Wolfgang Soergel
showed that the terraces of the river Ulm demonstrated all the most significant swings of temperature shown in Milanković's "insolation curves."
Soergel wrote enthusiastically, "It is seldom that two independent attempts,
based on completely different facts, bring the same results. I consider the
divisions and chronology of the ice ages to have been astronomically proven
and the problem solved." Another German geologist, Barthel Eberl, who
published a work 440 pages long based on studies of the terrain near the
Danube east of Regensburg and south of Ulm, said he "could not believe his
eyes" when he saw how closely his discoveries coincided with Milanković's
theoretical findings. Eberl's diagrams were developed from extensive exploration; Milanković's had come from work done at his desk. Eberl intended
to probe even further into the distant past, and Milanković extended his
calculations back to one million years ago on Earth.

In 1937, Milanković was fifty-eight years old and had been working for
twenty-five years on his astronomical theory of climates. According to his
memoirs, with the publication that year of his paper "Astronomical Means
of Examination of the Climate of the Earth's Past" in the *Handbuch der
Geophysik*, he was calling an end to his work in that field. It wasn't really
the end. Until Germany invaded Yugoslavia in April 1941, he continued to
refine his calculations, give them a more modern mathematical treatment,
and translate the most important part of his work into German. In 1938,
he published the final version of his astronomical theory, identifying the
variations in summer radiation at high latitudes in both the northern and
southern hemispheres as the primary cause of ice ages. These variations
were the result mainly of changes in axial tilt but were also affected by the
precession of the equinoxes. Taking into account changes in the reflective
power of the earth, he calculated how the geographic positions of the margins of ice sheets varied over the past million years.

Heavy German bombardment of Belgrade began on April 6, 1941. The Academy where Milanković taught suffered a direct hit, and the building where printers were a few days away from producing the German translation of his book *Canon of Insolation and the Ice-Age Problem* was reduced to rubble. A few days later, the Germans took Belgrade. Milanković and Tinka boarded up the windows of their house, dug a rubbish pit in the garden, and began a regimen of collecting water in cans from a faucet half a mile away and chopping wood to fuel an ancient cooking stove. They settled down to endure the occupation, struggle with food shortages, shelter Jewish friends, and help those friends acquire false documents in order to escape. On the brighter side, their son Vasko avoided being taken as a prisoner of war; and Milanković's translation had survived after all, suffering only water damage. In the autumn of 1941, his country's conquerors were able to read his book in their own language.

Milanković was sure that Hitler would eventually lose the war. "We knew who would win and who would lose, what we did not know was when. Therefore we had to survive until that unknown moment." Nevertheless, he found himself not hating his enemies and thinking less of himself for it. Referring to a German officer who was quartered in a room in their house, he commented, "They have killed thousands of us, and I talk to him without hatred. That I never hated anyone is abnormal. It is a weakness of my character." It was during the German occupation that Milanković decided he would write a history of science and his memoirs, so that no one else after his death would write a biography with "invented or incorrect facts in order to maintain the interest of the reader!" His history of science, beginning with Pythagoras and ending with Charles Darwin, kept him "in elite company throughout the war."

When the war ended, Yugoslavia became a communist country. Milanković's son Vasko and his wife Vera escaped and eventually moved to Australia. Sadly, they never again visited Yugoslavia, but they did maintain a steady correspondence. Tinka longed to visit them, but Milutin insisted that they had become too old for such travel.

During the 1930s and 1940s, geological work and discoveries seemed to confirm Milanković's theory, and most European geologists were won over to his camp. American geologists were more skeptical, and it was from those scholars that his theory began to be seriously challenged in the early 1950s.

Between 1946 and 1949, Willard F. Libby of the University of Chicago had developed the radiocarbon dating method, and this was almost immediately put to use in a worldwide effort to date the last ice age. At first the findings seemed consistent with Milanković's theory, but then a new discovery dealt it what looked to be a mortal blow. The new geological evidence came in the form of a twenty-five-thousand-year-old peat layer in Farmdale, Illinois, a deposit that could have formed only when the climate was relatively warm. Unfortunately, the Milanković theory showed twenty-five thousand years ago as a time that should have been very cold. Similar deposits were found in Canada and in Europe.

As radiocarbon dating continued, geological finds and the astronomical theory seemed more and more out of sync. Milanković himself was largely unperturbed by the rise and fall of interest in his theory, the various studies and discoveries that seemed, in turn, to corroborate it or disprove it, or the opinions of other people. He was certain his theory was correct and hardly bothered at all to defend it. "Every scientific achievement has its absolute value which will sooner or later be recognized—posterity being its high court."

When Milanković died in 1958, that court had not spoken. It was still very much up in the air whether scientists would eventually accept his theory or whether it would be relegated to the dustbin as a valiant but flawed attempt. By the mid-1960s, when radiocarbon evidence against the theory had continued to pile up, it had lost almost all its supporters. By the late sixties, it seemed completely out of the running.

In the mid-1950s, however, when most geologists were beginning to reject Milanković's theory, a group of researchers—most notably John Erikson and Cesare Emiliani, and, a little later, John Imbrie—had been focusing their attention not on land but on the sea bottoms. Taking core samples there and using them to interpret the climates of geological periods far in the past, they had found what seemed to some a rather convincing correspondence with Milanković's graph of radiation curves.

In 1965, seven years after Milanković's death, Wallace Broecker of Columbia University reported using more advanced dating techniques based on radioactive isotopes of thorium to study the fluctuations in the history of sea levels. He discovered that three dates of high sea levels—which would correspond with warmer periods—were a reasonably good match for Milanković's radiation curves for sixty-five degrees north: 120,000 years ago, 80,000 years ago, and the present. But soon more studies by Broecker, by Robley Matthews of Brown University and graduate student Kenneth Mesolella (who dated reefs off Barbados), and by others in Hawaii and New Guinea yielded a fresh puzzle: evidence of high sea levels during a period 105,000 years ago. Milanković's graphs had shown no evidence that that should be a warm period. Then Broecker realized that if he examined Milanković's curves for forty-five degrees north, rather than sixty-five degrees north, he found a distinct peak for 105,000 years ago. At lower latitudes, it is the precession cycle that has the strongest influence. Milanković's astronomical theory *could* account for high sea levels that evidence showed had occurred 82,000, 105,000, and 125,000 years ago.

A milestone in the way of better understanding of the three astronomical parameters that drive long-term variations of climate came in the 1970s from Belgian André Berger, with his discovery that the lengths of the precession cycles and the eccentricity cycles are not nearly so consistent as Milanković had thought. Berger's most recent numbers (2013)

A NASA satellite captured this image of the Scandinavian Peninsula in February 2003, showing jagged inlets and fjords—a landscape largely shaped by glaciers during the ice ages.

show that changes in axial tilt do take place in a faithful 41,000-year cycle, but the precession cycles and the eccentricity cycles vary considerably, with the most significant periods being for the precession cycle 19,000 and 23,000 years, and for the eccentricity cycle 100,000 and 400,000 years.

Evidence continued to weigh in on the side of the astronomical theory. Particularly significant, in 1976 James Hays, John Imbrie, and my friend Nick Shackleton—who first introduced me to the story of Milanković—had been analyzing cores taken in the Indian Ocean and established that significant changes in Earth's climate have coincided with changes in Earth's axial tilt and the precession of the equinoxes for 500,000 years. Milanković's theory came back into fashion as astronomers refined calculations of orbital history and geologists studied new evidence of climate history, and as improved dating methods emerged. It was soon evident that at least as far back as Permian times (a period lasting from 299 to 251 million years ago), major variations in Earth's climate could be attributed to changes in how much radiation reached Earth, which can in turn be attributed to changes in precession, axial tilt, and the eccentricity of the orbit.

In the last decades of the twentieth century, many researchers were beginning to suspect that it would have taken more than the cycles revealed by Milanković and the changes in the radiation balance linked with them to cause the enormous, dramatic shifts between glacial and interglacial periods in Earth's history. Perhaps the picture was not as simple as Milanković had thought. The suggestion was that Milanković's cycles form a background influence that is essential in producing ice ages but aren't powerful enough by themselves to explain them. Within Earth's own environmental systems, other factors must have operated to amplify the effect of the cycles. Fluctuation in ocean currents and airflow currents such as the jet stream and the El Niño winds have played a role, and other suspects are volcanic eruptions; the movement of tectonic plates and their results (such as the Tibetan uplift); the dryness or wetness of different parts of the earth in some eras; and carbon dioxide levels—even long before human beings had any influence on them. None of these factors works in isolation, and

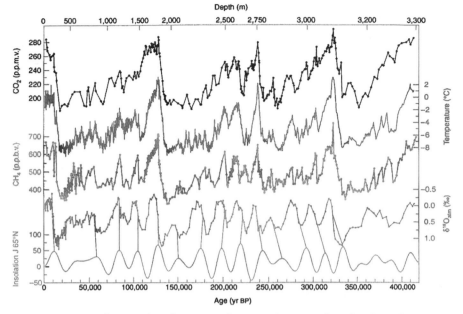

420,000 years of ice core data from a Vostok, Antarctica research station. From bottom to top, the successive lines represent solar variation at sixty-five degrees north due to Milankovitch cycles—readings having to do with isotopes of oxygen, levels of methane, relative temperature, and levels of carbon dioxide.

the interaction, feedback, and balance among them—and between them and the astronomical cycles—are extremely difficult to calculate.

It remains to be seen—though not by anyone alive today—where the elaborate choreography of Earth's orbit and other suspected influences will lead Earth's climate in the future. Milanković's calculations predict that Earth will eventually enter another ice age, but there is an element that he was unable to consider in his equations—the actions, inactions, and interventions of human beings. Does our use of fossil fuels and our production of other greenhouse gases heat the earth sufficiently to overcome the powerful astronomical forces he revealed? Will we run out of those fuels or decide to curtail their use before irrevocable damage is done to the progression of natural cycles? Are we *capable* of irrevocable damage, or will nature and physical laws prevail in the end? Regarding that, the high court has not yet spoken.

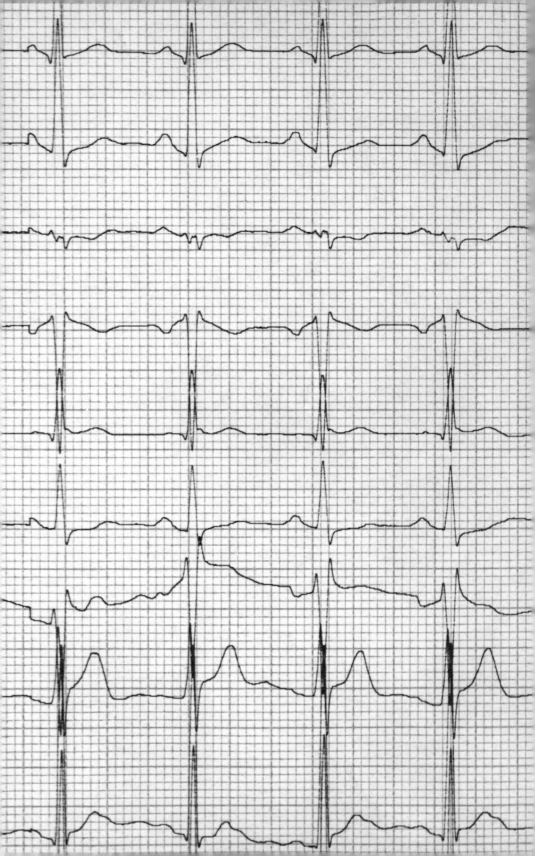

Barry Sterman

THE ELUSIVE QUALITY
OF STILLNESS

(late 1950s–present)

n the history of science and medicine, there have been break-throughs that occurred too soon, before scientific knowledge was in place that would have allowed researchers to fit what seemed hardly more than scientific oddities into mainstream science. There are other cases where enthusiastic, sometimes "amateur" follow-up on a discovery carried it so far afield from where more systematic, responsible research ought to have taken it that its first simplicity, attention to detail, and credibility were all but lost. Barry Sterman's surprise discovery in 1965 of neurofeedback falls into both categories.

In the late 1950s Sterman was in his first year of graduate school, studying psychology at the University of California, Los Angeles. He was assigned to tutor an intelligent eleventh-grade boy who was performing in school at only an eighth-grade level. The boy had a strange, stiff way of walking, an unusually pale complexion, and an inability to show emotion or communicate beyond a few one-syllable words. He'd been diagnosed as possibly schizophrenic and borderline delusional. Sterman was studying for a midterm exam on material that included thyroid function "and all of this information was running around in my head," he remembers. "So I came from class to work with this boy, took a look at him and said to myself, 'My God, he's a walking definition of hypothyroidism.'" That off-the-cuff student diagnosis pointed the boy's parents in the right direction to find help, and thyroid medication saved what otherwise might have been a tragically doomed young man. "I realized that I would have to study biology

An electroencephalogram, or EEG recording.

and physiology as well as psychology if I was going to work with human beings." The physiology of our brains and bodies, Sterman concluded then, are inextricably interwoven with our thoughts and emotions. Body and mind can't be studied separately. That realization set the course for his future. He earned a PhD in neuroscience and psychology from UCLA in

Portrait of Barry Sterman.

1963 and after that went on to postdoctoral work as a research scientist in neuropsychology research at the Veterans Administration Medical Center in West Haven, Connecticut.

By the mid-1960s, Sterman had moved to the Veterans Administration Hospital in Sepulveda, California, where he was a research psychologist. During this time, he was also an assistant professor in the department of anatomy at the UCLA School of Medicine. His principal interest was investigating brain activity during sleep, and his subjects were cats and monkeys. Sterman decided to follow up on Ivan Pavlov's famous "Pavlov's Dog" experiment, taking advantage of a modern advance that Pavlov would have welcomed: the electroencephalogram, or EEG. With an EEG, Sterman could study the electrical activity in an animal's brain. Sterman was asking whether, when we fall asleep, it is because we choose to do so or because it happens automatically. Pavlov's dog, when faced with confusing signals, promptly fell asleep, and Pavlov had concluded that it is possible to fall asleep deliberately. Sterman hoped to find out whether or not Pavlov was right.

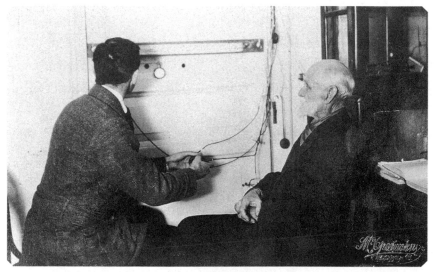

Ivan Pavlov (right) conducting one of his stimulus-response dog experiments, c. 1921; the operator, left, is observing the dog through a periscope in the door.

Sterman brought thirty cats into his lab and deprived them of food long enough to make them hungry. He selected a cat, placed electrodes connected to an EEG machine on its head, and moved it to an experimental chamber where there was a lever and an empty bowl. If the cat pressed the lever, the bowl would fill with a luscious mixture of chicken broth and milk. The cats were eager to go through this procedure. They were, after all, hungry, and this was more interesting and not a great deal more challenging than walking up to a bowl and eating. Sterman introduced a new wrinkle: A tone sounded in the chamber. When the tone was sounding, pressure on the lever produced nothing. The cats learned to wait for the tone to stop before pressing the lever, and this waiting had a special quality. Sterman found that the cats entered what seemed to be the same state of consciousness that an animal exhibits when waiting to pounce on its prey—completely still, very alert. During this stillness, the cat's EEG showed an electrical frequency pattern that no researcher had studied in detail before: "a clear rhythmic change. It was fascinating . . . it didn't exist in the literature of the time." It was a specific rhythmic signal peaking in the range of twelve to fifteen hertz (see pages 270–271); a "beta" signal over a part of the brain known as the sensorimotor cortex. Sterman dubbed it the "sensorimotor rhythm," or SMR. He wondered whether a cat might learn to produce that specific range of brain waves at will, and, to find out, he modified the experiment again. Without any lever or tone, the broth and milk would now appear whenever the cat produced a half-second of that frequency. The cats learned to produce SMR at will, even more quickly than they had learned to press the lever for food. Behavior, Sterman had found, could be altered by EEG conditioning. Clearly, the training had altered neural activity in the brains of these cats.

SENSORIMOTOR CORTEX

The *sensorimotor cortex* is the part of the brain that receives sensory information from our bodies and our environment through our senses—vision, hearing, smell, taste, touch, the inner ear (our sense of balance), and proprioception (awareness of the relative position our body parts and the amount of effort needed in movement)—and that initiates a response (motor output).

Sterman and his assistants also studied the EEG of the cats during sleep. This, too, had been altered by the training. The cats slept much more soundly and woke up many fewer times during sleep periods. When Sterman repeated the study with rhesus monkeys (he offered them Spanish peanuts instead of milk and broth), he got the same results. Sterman published all of this in the medical journal *Brain Research* in 1967. It had been an extremely interesting but obscure little experiment, and Sterman had been the first to isolate the twelve to fifteen hertz frequency and give it a name. But his findings seemed to have no significant practical relevance to science or medicine, nor any clinical application. Or so Sterman and his associates thought.

The human brain is a storm of electricity as brain cells—neurons—receive and transmit information in a network of trillions of connections. Each neuron receives impulses from other neurons through fibers called *dendrites* and sends outgoing impulses through a fiber called the *axon*. A neuron has many dendrites but only one axon. The end of an axon doesn't *quite* touch the dendrite or cell to which it is sending a signal. Instead, when an electrical impulse comes to the end of the axon, it triggers the release of a neurotransmitter that jumps the gap. Working in this way, each neuron communicates with others—both nearby and all the way on the other side of the brain. The brain's networks are made up of an inconceivable number of connections and arrangements, and even when we think of the brain as at rest, waves of electrical activity are sweeping across it. It would seem that the electrical potential of the brain is immense, but in fact it is a tiny fifty millionths of a volt and difficult to measure. For comparison, the voltage measured from the heart is ten times that great.

The knowledge that electricity is involved in the workings of the brain and can be measured by instruments was not new to science in the mid-twentieth century. In the 1870s and 1880s, Richard Caton, a doctor in Liverpool, England, studied the brains of rabbits and monkeys using a device known as a galvanometer. The wire and coil of a galvanometer vibrate when electricity, even in minute amounts, is detected. Caton attached a small mirror to the coil and set up a bright lamp to cast a narrow beam of light onto the mirror. The mirror reflected the beam onto an eight-foot (2.4-meter)-high scale painted on a wall. The stronger the electric signal, the higher up on the scale the light shone. Caton attached electrodes from this instrument to the brains of rabbits and monkeys and found that when an animal chewed its food, moved, or had a light directed into its eyes, there was a spike in the height of the beam of light. Even across the unopened skull of an animal, he was able to detect a weak flow of electricity. This experiment and its result were the early ancestors of the EEG.

In 1924, German psychiatrist Hans Berger invented electroencephalography or EEG. Speaking at the 2002 meeting of the International Society for the History of the Neurosciences (ISHN), modern neurologist David Millett called that invention "one of the most surprising, remarkable, and momentous developments in the history of clinical neurology." As a young man in the German military, Berger had an accident that nearly cost him his life and gave him a lifelong belief in the reality of mental telepathy. He

A moving coil galvanometer. The wire and coil of a galvanometer vibrate when any electricity is detected.

was convinced that he had transmitted his thoughts over a distance, and he was determined to find out, scientifically, how this could have happened. His own eccentric, "thermodynamic" model of "cerebral energetics" motivated a rigorous, lengthy research program. Berger speculated that if it were possible to arrive at a good estimate of how much energy the cortex (the brain's outer layer) stores and to measure precisely how much heat and electricity were being produced by cortical tissue, he could calculate how much energy was being converted into thought, feelings, and emotions—psychic energy. At a later stage in what became a three-decade, unlikely investigation hampered by almost debilitating bouts of depression and self-doubt (he had little training and was operating delicate instruments at levels near the borders of their sensitivity), Berger tried to measure the energy converted into heat and electricity during different mental tasks. In order to record electrical currents of the brain from the scalp surface of a conscious subject—his fifteen-year-old son Klaus—he invented what he called his *Hirnspiegel*, or brain mirror. We call it EEG.

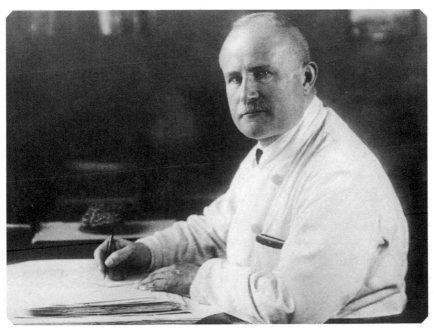

Photograph, c. 1924, of German psychiatrist Hans Berger, inventor of electroencephalography.

Berger's device, like Caton's, utilized a beam of light that vibrated with the electrical signal. Instead of directing the beam to a scale on the wall, Berger directed it onto a moving piece of photographic paper, thus creating a graph of the brain wave. With this device, he was able to record a rhythmic disturbance having a frequency of about ten hertz (the "ten-hertz range"). These waves became known as Berger's "alpha rhythm." In 1929, Berger announced and described his technique for recording the electrical activity of the human brain from the surface of the head. The idea was met with such incredulity and skepticism from the scientific and medical establishment that the discovery was not accepted until Edgar Douglas Adrian—better known as Lord Adrian, professor of physiology and Nobel laureate—replicated it at the University of Cambridge in 1934. Berger's life nevertheless ended tragically. In 1938, when Nazi officials ordered him to fire his Jewish staff members, he refused. He was forced to retire. With the loss of opportunity to continue his research, and a precipitous decline in his health, his depression worsened. In 1941, Berger committed suicide.

HERTZ RANGE

What is meant by the "ten-hertz range" that Berger discovered and the "twelve to fifteen-hertz range" in which Sterman would later work? The number of "hertz" used in measuring electrical activity indicates how many cycles there are per second. One hertz is one cycle per second. Ten hertz obviously means the cycles are happening more frequently than with five hertz, and so forth. The number of hertz is the "frequency" of cycles—basically, how often they happen. We use the same term to refer to the electrical current in our homes. In the United States the house electrical supply is at 60 hertz. That means the current changes direction or polarity 120 times, or 60 cycles, per second. Broadcast transmission is at much higher frequency rates and is usually expressed in kilohertz or megahertz. The human brain operates normally in a range of one to forty hertz. Neuroscientists group the frequencies into categories. *Delta* is the 1–4 hertz range and occurs during sleep and in some comas. *Theta* is the 4–8 hertz range, when one is conscious but in a sort of twilight state, between deeply relaxed and sleeping. *Alpha* is 8–12 hertz, a relaxed but awake state. Hans Berger's 10-hertz discovery was in this range. *Beta* is 13–circa 30 hertz—the range of normal, awake consciousness. Sterman's discovery, the rhythmic signal peaking in the range of 12–15 hertz, which he calls SMR, is a beta signal.

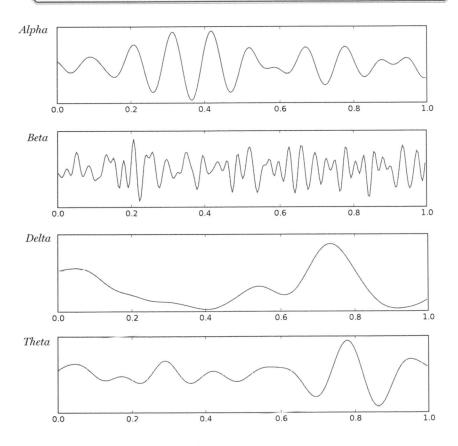

Diagrams showing alpha, beta, delta, and theta waves.

In the post-war years, EEG rapidly gained widespread recognition by eminent researchers and was utilized increasingly in diagnosis in the United States, Britain, and France. It was this technology—greatly refined over the years from Berger's crude machine—that Sterman used to study the brains of his cats and monkeys, and he would soon use it in another study with more practical relevance. Not long after Sterman's cat experiments, he received a phone call from a friend, David Fairchild. NASA had contracted Fairchild and drug researcher Gordon Allies, the inventor of amphetamine, to undertake research on a highly toxic rocket fuel known as monomethylhydrazine (MMH). MMH was apparently causing disturbing symptoms in workers who came in contact with it or breathed its fumes—nausea and severe, sometimes deadly, epileptic seizures. It was feared that even lower exposure over extended periods could cause hallucinations or disrupt the ability to think clearly. The situation had become life-threatening for astronauts whose survival in space depended on high cognitive function. When Mercury astronauts flying over the Pacific Ocean thought they saw natives waving at them, it was clear that some investigation was in order—and not in the Pacific Ocean. Sadly, Gordon Allies died when testing on himself a chemical compound he was developing. Fairchild invited Sterman to replace Allies. Sterman agreed.

Sterman brought in fifty cats and injected them with the rocket fuel. After a few minutes, the cats vomited, salivated, panted, and made strange noises. After an hour, all except ten lapsed into grand mal epileptic seizures. In seven of the remaining ten, the onset of seizures was greatly delayed, and three had no seizures at all. What was different about those ten cats? "We couldn't figure out what was going on," Sterman said. "Answering that question defined the next ten years of my research life. I shifted my research focus to EEG indications of brain function and began looking for ways to study them. The roots of brain modulation training began with that effort."

Going back to the cats' medical records, Sterman found that he had used those ten cats in his previous EEG conditioning experiment with the lever and the tone. Sterman concluded that the ten cats had learned how to produce SMR in the earlier experiment, strengthening their brain function and heightening the threshold at which seizures would occur. More specifically, their brains had become able to prevent—or at least resist, to a certain extent—the spread of slow theta waves across the motor cortex, which are the cause of seizures. They had learned to fight off epileptic seizures.

Sterman was and is particularly eager to have it understood that the effects of SMR neurofeedback are not a placebo. They are physiological; something really is changing in the mechanics of the brain. A placebo effect is a common phenomenon in the testing of pharmaceuticals and other medical testing and experiments, but there is no possibility of a placebo effect in cats or monkeys, "who were in no way influenced by expectation," Sterman points out.

Clearly the next question to tackle was whether or not SMR neurofeedback could work for human sufferers of epilepsy, either to reduce the number of seizures, or to prevent them entirely. "It couldn't be too hard for humans to learn this when cats did it in two weeks," Sterman remembers thinking. Did human beings have the SMR signature rhythm? At this time, there were still doubts among neuroscientists about whether EEG was sensitive enough to pick up brain waves through a human skull. An opportunity came Sterman's way to test the procedure on the brains of several cancer patients who had had part of the skull bone removed. Without the interference of an intact human skull, Sterman got beautifully clear readings of the frequencies, and SMR in humans was confirmed. These patients had, of course, not been epileptics, so it remained to be seen whether human beings could alter their brains as the cats had done, with the same positive results.

Sterman's first human subject was Mary Fairbanks, a computer coder working for one of his colleagues. Fairbanks was twenty-three and had suffered since age eight from a major motor disorder that brought on severe grand mal seizures two or more times a month. Drugs had not helped. She had kept a carefully detailed journal of her seizures through the years, recording when they happened and how serious they were. The National Institutes of Health (NIH) and researchers at the University of Wisconsin also had thorough records of her condition. This scrupulously detailed "before" picture was essential to making her treatment a meaningful experiment.

When someone is in a normal state of consciousness, his or her brain is operating in the beta range, usually from 13 to 18 hertz (refer to pages 270–271). In someone suffering from epilepsy, a part of the brain isn't stable and is unable to resist an invasion of slower theta waves (in the 4 to 8 hertz range) from spreading across the motor cortex. Soon other areas of the brain are also producing the lower frequency, and this disrupts normal motor function. The goal of Sterman's work with Mary Fairbanks was to train her brain to resist this spread of theta waves. If his findings with the cats and monkeys held up for humans, that resistance could be strengthened. Seizures would happen less frequently and maybe be eliminated entirely.

Sterman first attached Fairbanks to a neurofeedback apparatus in 1971. The goal was to have her produce a beta rhythm signal in the 12 to 15 hertz range over the sensorimotor cortex (refer to page 266)—Sterman's "sensorimotor rhythm" or SMR. One might hope that, in the spirit of the cat experiment, she would have been offered chocolate cake whenever she succeeded . . . but no. When she was producing SMR, a green light went on. When she was not, the light was red. Her task was to keep the green light shining, encouraging the higher frequency waves and discouraging the lower frequency. There is nothing particularly unusual about an SMR frequency. We all pass through it often, spending a split second in it now and then. Sterman hoped that Fairbanks's training would allow her to stay in that frequency for longer intervals. Her cortex would be learning how to maintain stability.

For three months, Fairbanks trained twice a week for one hour each time. At the end of that period, she was seizure-free. The recovery lasted, and eventually she was able to get her driver's license—an impossible feat for someone subject to seizures. She became a more outgoing, confident person. In 1972, Sterman reported the case in the journal *EEG and Clinical Neurophysiology*, writing that "training of this sensorimotor rhythm resulted in a striking enhancement of the rhythm's occurrence . . . and a marked suppression of seizures. Changes in sleep patterns and personality were noted also."

What actually was going on in the brain with these changes of frequency? Early on, Sterman postulated that the mechanism being affected by EEG training was the mechanism by which the part of the brain called the thalamus (which relays motor and sensory signals to the cerebral cortex) regulates and stabilizes the cortex, which is by its very nature hyperexcitable. This regulating and stabilizing mechanism operates rhythmically in the 12 to 19 hertz range. The thalamus works by projecting to localized regions of the brain and at the same time to broad regions of the cortex (see image below). While EEG experts using neurofeedback reward specific brainwave frequencies (the selection of the frequency bands depending on the specific goals of treatment), clinicians, guided by what neuropsychologists and neuroanatomists have learned about where different functions are localized in the brain, vary the placement of electrodes on the skull, with both very specific and more general effects.

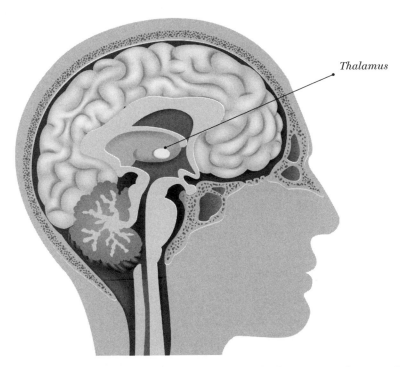

Cross section of the human brain; the thalamus—which relays motor and sensory signals to the cerebral cortex—is at the center.

In the 1970s, Sterman was moving from success to success. Other important researchers were eager to come to California to work with him. Laboratories around the world were repeating his results. It was increasingly easy to get funding for his work. In the next phase of his research, he treated four epileptics and achieved a 65 percent reduction in grand mal seizures. Sterman published those results in 1974 in the journal *Epilepsia*. In 1976, grants from the National Institute of Neurological Disorders and Stroke—a branch of the NIH—allowed him to schedule tests using eight subjects and make the testing more rigorous.

This experiment had an A-B-A design, a mode of testing that has since been abandoned on the grounds that it is too manipulative and potentially harmful, even if the end result is good health. Sterman's patients first trained for three months in the same way Sterman had trained those in his earlier study. They learned to increase their SMR waves and suppress the theta waves causing their seizures. As hoped and predicted, the number of their seizures dropped significantly. Without letting them know that he was making a change, in the B stage of the experiment Sterman began rewarding his patients when they *encouraged* the detrimental theta waves and *suppressed* SMR. The number of seizures returned to where it had been at the beginning of the experiment. After an additional three months and again without the patients' knowledge, the protocol was reversed a second time. The seizures were reduced again. This was a rigorously controlled and imaginative experiment, considered ethical at the time, and in this case, it did no one any harm. It certainly supported arguments for the effectiveness of neurofeedback. Again, Sterman published his findings, reporting that six of the eight patients had significant and sustained seizure reductions. Overall, the result was a 74 percent reduction.

With those results to back him up, Sterman obtained another NIH grant. This time the study was to last three years, and he would work with twenty-four patients. He was particularly interested in how well his treatment would work for people whose epilepsy had not responded to drugs. His subjects included several who were on a waiting list for surgery called an "anterior temporal lobectomy," which removes a small area of brain tissue that is the source of the spread of the low frequency that causes seizures. In a complicated study, Sterman divided the twenty-four patients into three groups, two of which were control groups. Members of Group One, a control group, received no treatment but were taught how to record their seizures and keep logs. In what is known as a "yoked study," the members of Groups Two and Three were paired off, with one member of each pair being from Group Two and the other from Group Three. One member of each pair was given real neurofeedback. The other member was not but was led to believe that he was because he was hooked up to the results of his partner. In other words, Jane (Group Three) was responding to a recording of Sue's (Group Two) EEG signal—not to her own. If Jane, who was receiving no real neurofeedback, began to reduce her seizures, then Sterman would know that a "placebo" effect was at work.

To Sterman's satisfaction, those receiving real neurofeedback (such as Sue in Group Two) did reduce the number of their seizures—in fact, several Group Two patients ended up completely free of seizures. However, those with the "sham" treatment (Jane in Group Three) did not. Overall, there was a 60 percent drop in the average rate of severe seizures. When the study was finished, those in the two control groups were also given real neurofeedback training. Sterman's work and his results were replicated in several independent studies in other labs. His yoked study had proven beyond reasonable doubt that neurofeedback was working for the treatment of epilepsy.

During the 1970s, some medical people were beginning to use neurofeedback in actual clinical treatment of patients, and Sterman knew it was time to find out, more systematically, how effectively it could be used in a hospital or doctor's office—and also how well it would work for longterm management. In 1982 he received a NIH three-year grant of $70,000 a year.

As the first year of the grant was coming to an end, a letter arrived from the NIH. A committee reviewing his grant asked for a double-blind aspect to his work. Sterman was surprised and puzzled. It was almost unheard of for a grant committee to ask for such changes when a previously approved project was well under way and proceeding as planned. Sterman was preparing to comply with their wishes when a second letter arrived telling him that his work had achieved its objective and no further research was needed. His funding was cancelled.

Over three decades later, Sterman is still bitter about that decision, which he attributes to politics at the NIH. In the late 1990s he told interviewer Jim Robbins, author of *A Symphony in the Brain*,

> *Doctors want Ph.D.s [sic] to be in the laboratory documenting procedures, documenting drugs, documenting the treatments they could apply. And here I was proposing a protocol for a long-term treatment of epileptics. It was a turf battle, in my opinion. Pure politics.*

In an e-mail to me, Sterman added, "This opinion was later validated by a colleague who had been on the committee!" As the 1970s progressed, pharmaceuticals were taking over the field of health from other forms of treatment. Whatever its motivation, the NIH decision to cancel Sterman's funding was a deathblow to a promising field of medicine.

Since then, Sterman has published more than one hundred fifty papers in prestigious journals and applied for numerous grants, hoping to return to his initial research. "But the [NIH] will not give us grants. We've written solid grants but the minute you use the terms EEG feedback or neurofeedback certain people's minds snap shut." As far as its use in mainstream medicine was concerned, in 1983, neurofeedback became a lost science.

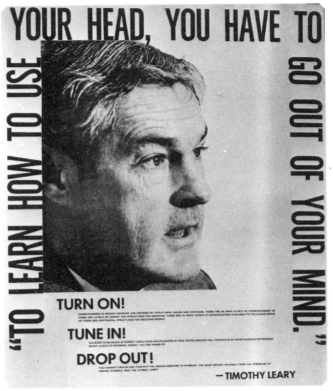

A poster featuring psychologist and counterculture figure Timothy Leary, who advised expanding levels of consciousness with LSD to "Turn On! Tune In! Drop Out!" The movement of which Leary was a part greatly affected the credibility of Sterman's field of science.

Something in the culture of the time was also working to Sterman's disadvantage: the cult rage in the sixties, seventies, and early eighties for "alpha-training" and popular efforts at "biofeedback."* This was the age of the Beatles, of the rage for transcendental meditation, Woodstock, free love, LSD, Timothy Leary. Academic meetings that in the past had attracted conservative neurofeedback and biofeedback professionals cautious about making unverified claims for their fields were suddenly drawing others with no basic knowledge, who had wild and poorly founded hopes. These enthusiasts, some with the long hair,

---

* Biofeedback is unlike neurofeedback in that it is not limited to activity in the brain but involves the training of other body functions and other parts of the body *including* the brain.

bare feet, and flowing robes that characterized the era, were convinced that altered states of consciousness could solve many societal problems. While confidence in a strong future for neurofeedback was not lacking among more conservative researchers, the highly emotional, much-too-speculative claims from those with no scholarly knowledge or training caused serious university neuroscience researchers, Sterman among them, to draw back in fear that they would lose their own credibility. Sterman was careful to distinguish his work from this fad, to publish only in highly respected journals and to conduct his investigations in a rigorous scientific manner that no one could fault, but anything that seemed related to the popular version of EEG biofeedback—as his work with SMR training appeared to be—began to be treated with suspicion and outright distain in medical research and in the medical profession. Sterman's results were not derogated, but they were relegated to the status of a scientific oddity almost entirely off the radar screen of serious science and medicine. The stigma would last for decades and, to this day, has not disappeared.

In the mid-1980s, with no funding to continue research that would provide more data on the effectiveness of neurofeedback, refine his own knowledge, and decide on neurofeedback's potential usefulness, Sterman took his files on SMR training in cats, monkeys, and humans and stored them away. He was still able to make a living. The U.S. Air Force hired him to study how pilots pay attention when they are flying and how cockpits might be designed for better efficiency. On a part-time basis, he eventually used neurofeedback to treat a few epileptic patients at Hollywood Presbyterian Hospital. But when it came to the serious, systematic, adequately funded testing of neurofeedback that he had once planned, he was effectively benched for a lifetime.

Back in the 1970s, Joel Lubar of the University of Tennessee, Knoxville, had begun to use neurofeedback to treat a different condition. Lubar was studying ADD (attention-deficit disorder; reclassified as ADHD by the Diagnostic and Statistical Manual of Mental Disorders [DSM] in 1987 and referred to as such herein) in children and adults when in 1972 he first came across papers that Sterman had published about his results with epilepsy.

> *Sterman's paper said there is a brain wave frequency known as SMR that may be dysfunctional in epileptics, and when that rhythm is partially restored, a resistance to seizures occurs. As soon as I read that, I said, "My God, I think this will work for controlling hyperactivity in children because the circuitry is very similar." If you could quiet the motor responses for epileptics, then quieting the motor responses in hyperactive children should be a piece of cake.*

Lubar first replicated Sterman's epilepsy results in his own lab and then went to California to work for a year under Sterman. The most telling investigation he did during this period involved four children diagnosed with ADHD. The study was an A-B-A design similar to the one Sterman used with the eight epilepsy patients. He trained the children using neurofeedback until their symptoms disappeared, then reversed the training until they reappeared, then reversed the training *again* until the symptoms disappeared once more. It was possible to treat ADHD successfully without drugs.

Before the mid-1980s, both Sterman and Lubar (and no one else) had been working in the field of brainwave training on the rigorous, scholarly level at which Sterman had begun. After Sterman's research funds were cut off, Lubar alone was able to continue with some systematic testing, but other than that, methodical scientific investigation came to an end. Sterman's and Lubar's ideas and methods became increasingly difficult to "sell" in the world of establishment scientific research, though there were a few odd moments of public attention. In the spring of 1991, *Woman's Day* magazine published an article about Lubar's success with ADHD–diagnosed children, reporting that neurofeedback had made the children less impulsive, made them able to concentrate for longer periods of time, and even raised their IQs ten to fifteen points.

Meanwhile, others, practicing on their own rather than in universities, research hospitals, or research laboratories, concentrated almost entirely on using different versions of neurofeedback in the ad hoc treatment of patients, rather than on testing its use and results systematically in controlled experiments. By the end of the twentieth century, the neurofeedback field was no longer tiny, but it was disorganized, somewhat cantankerous, producing only anecdotal evidence. Some were thinking of Sterman's legacy as a proliferation rich with possibilities, others as a morass of confusion and disagreement with scientifically unsupportable fringes. Pushing the envelope was not unfailingly good science. Amateurs and others whose academic credentials were minimal were taking the field into areas that stalwarts like Lubar and Sterman felt were too speculative. The evidence for the efficacy of some treatments was not well enough corroborated to serve as anything other than encouragement for someone else to pick up the ball and test the possibility more carefully. Usually no one did.

Neurofeedback and its offspring, as a field, were no longer rigorous science. Too many claims, too many competing methods, too many ways of combining neurofeedback with other modes of treatment, too many untrained practitioners, too many pieces of equipment available for anyone to purchase, too many patients handling their own cases at home with little supervision. . . . Anyone seeking treatment had no way to make an educated choice of where to get it. General practitioners and specialists dealing with brain-related illnesses were unlikely to recommend anything in this field and were entirely unequipped to guide a patient to the most appropriate version of treatment. The clear-cut, well-substantiated results of Sterman's early testing of neurofeedback—with their promise of a new field of effective treatment that would be respected and widely used by physicians and hospitals—had stalled when his funding was withdrawn. Instead, a highly promising discovery had birthed a quagmire. That is not to say that there were not and are not practitioners of neurofeedback performing clinical work and pursuing a deeper understanding of their subject in a serious, scholarly manner. However, each of them pretty much has to find his or her own way on an uncharted journey.

One common theme among these practitioners is the belief that there is a physiological connection between the different conditions and problems that appear to respond to neurofeedback training. On the surface of things, on the treatment level, the connections can't be missed: ADHD often seems to be attributable to trauma at birth, which can be a kind of head injury. Adults suffering from the results of head injury have symptoms similar to ADHD. ADHD often has such symptoms as bed wetting and night terrors, which are sleep problems and may also be related to seizures, because the majority of these occur at night. Other conditions that fall into the category of "underarousal" conditions also show up in sleep problems and cognitive deficits similar to those in ADHD. The neurological substrate that many think must underlie ADHD, epilepsy, problems with sleeping, depression, and deficits attributable to head injury has to lie at a very basic level in order to be affecting patients and treatment in all these categories. One of the criticisms leveled at the entire field is that the treatment of symptoms goes on without a basic understanding of the deeper physiological connections between these conditions and problems. With few exceptions, there has been much more interest in what neurofeedback can do and how practitioners can make it work than about what it is or *how* it works. At the same time, some careful research and observation at the clinical level combined with state-of-the art information from other fields of brain research have not completely failed to offer a window into the brain mechanics that underlie neurofeedback:

First, an EEG, as we've seen, measures the electrical activity in the brain, which is recorded by electrodes placed on the scalp of the patient. Each electrode covers about one hundred thousand neurons, which isn't many given the number of neurons in a brain (eighty to a hundred billion). As explained earlier in this chapter, each neuron has many dendrites that pick up information from other neurons. This information has the capacity either to excite the neuron, causing it to fire and pass the information on to other neurons, or to fail to excite the neuron to pass along the information. If the neuron fires, that firing will send an electrical impulse down along to the end of its axon, where it releases the neurotransmitter molecules that jump the gap to receivers on another neuron. With so many dendrites and neurons around, this one axon could communicate with as many as a million neurons, sending each of them a message to fire and pass the message along. Hence, in neurofeedback treatments, each of the hundred thousand neurons covered by an electrode is affecting, and is affected by, a huge number of others, either near it or much further away in the brain. The pathways and networks formed in this way influence areas all over the brain. Furthermore, it is well-known and well-documented that there are feedback loops between neurons in the cortex of the brain and the thalamus, which has important connections to the brain stem. Sterman postulated early on that the thalamus plays an essential role in neurofeedback. So when his patients managed to effect a change in their EEGs, that meant they are influencing feedback loops that connect throughout the brain. The effects of training at just one electrode site could be significant and occur very rapidly.

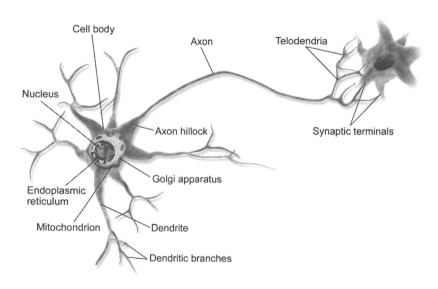

A multipolar neuron—the most common type of neuron in our nervous system—has a long axon and a crown of multiple branching dendrites.

Sterman comments now that even without controlled research specifically studying neurofeedback, there is much better understanding of the mechanisms that lead to the SMR pattern in the sensorimotor cortex (see page 266) and how this plays out when an individual is either awake or asleep. He explains that the mechanism that produces SMR in the brain is theoretically understood today as follows: The electrical events that give rise to the SMR EEG pattern originate from a group of neurons that form part of the thalamus called the ventrobasal nuclei, which relays messages having to do with the current state of muscles and tendons in the body. When an individual becomes quiet with no intention to move, cells in the nuclei that are specifically concerned with movement begin to shut down, setting up a general oscillation between other thalamic nuclei and cells with axons headed to areas in the cortex that are related to a bodily map of movement. This oscillation produces the EEG rhythm that Sterman called the Sensorimotor Rhythm or SMR. As an individual learns to control his or her SMR rhythm through EEG neurofeedback, the sustained quieting of movement can bring about a variety of changes in sensory processing. It can be associated with an increase in visual attention, showing up in a suppression of the so-called *alpha rhythm* in the visual areas of the cortex. Take driving, for instance. We all know the effect that adjusting knobs or searching for a fallen item has on us, even if we are also keeping our eyes on the road. Movement distracts visual processing. The SMR prevents this disruption and can facilitate sensory attention. This kind of reciprocal interaction between muscles and sensory activity, and even different muscles, provides for the exquisite control of attention and movement in humans and higher animals that Sterman observed in his cats. When an individual is awake, the SMR acts reciprocally with other sensory reception systems that are responding to stimuli from a person's internal or external environment, promoting a suppression of motor activity and facilitating attention and sensory perception. During sleep, where the SMR shows up in an EEG pattern called "sleep spindles," this motor inhibition reduces disruptions and stabilizes the state of sleep. In people whose attention or sleep quality is disturbed, these effects are therapeutic.

This current understanding of the mechanisms of SMR provides the underlying explanation of what is going on in the brain as a person—through the self-regulation of EEG activity using neurofeedback—develops the abilities to be more physically still and relaxed and to inhibit feelings of restlessness and fidgetiness. When motor function and sensory attention are mutually tuned down, a state detected by patterns of appropriated EEG rhythms, we can know that the stage is set for sleep.

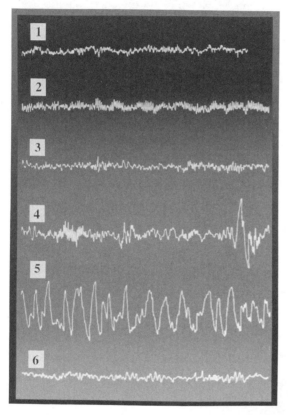

An EEG recording of brain waves in various states of sleep. Trace 1 shows a person whose brain is awake; in trace 2 (alpha waves), the person is awake but with closed eyes; in trace 3 (theta waves), the person is falling asleep; in trace 4, more complex patterns begin to emerge; and in trace 5, the standard delta-wave pattern usually associated with sleeping is shown. Trace 6 shows rapid-eye movement (REM), a period during sleep when the brain becomes unusually active and dreams occur; it looks similar to trace 1, when the person was awake.

Other normal rhythmic patterns in the EEG are produced in a similar way but show up at different frequencies based on differences in the pathways between the thalamus and the cortex, as well as on the arousal level.

In all this interplay, as neural networks in the brain vibrate across the entire EEG spectrum, constantly changing and doing so individually but with extremely complicated interconnections, the thalamus acts like the conductor of a large symphony orchestra, causing different frequency bands from different cortical areas to rise and fall, incorporating input from the cortex and from the individual's arousal level.

Broadly speaking, neurofeedback training today takes place at two levels of electrical frequencies: beta training, and SMR training. Treatments at these two levels have noticeably different effects. Beta training appears to affect conditions of underarousal, which can result from brain injury or disease or have genetic origins. SMR appears to affect conditions of overarousal, such as anxiety, hypervigilance, and higher susceptibility to stress. Taken together, the training appears to restore a normal arousal level as well as control of that level. Raising the seizure threshold with training (as in Sterman's experiment with the cats) suggests that the training makes the brain increasingly stable in the face of hyperexcitability in the cortex. The training appears to make the brain's basic arousal mechanisms, where they are deficient, more susceptible to self-regulation.

Experimentation at the treatment level, sometimes by unsupervised individuals who treat themselves in their own homes with equipment they have purchased, has produced a plethora of anecdotal evidence. It is no easy matter to judge what of this evidence is valuable and what is less so or even worthless. A few serious researchers in this under-controlled field are still using tried-and-true scientific criteria to make these judgments. The field of neurofeedback is arguably less dismissive of outside-the-box findings than many other fields of medicine. It is, after all, itself considered unacceptably outside-the-box by many people. In defense of anecdotal evidence, Siegfried Othmer, chief scientist at the EEG Institute in Woodland Hills, California, has written,

> Science in general, and in particular medical science, is not impressed by individual case histories, which are routinely dismissed as "anecdotal data." However, they can nevertheless be useful in the present context to calibrate . . . expectations of what this new technique may be used to accomplish. They are also useful scientifically, by forcing our attention on new

*phenomena that have simply not yet been studied extensively,
but point the way to the future. Science must start somewhere,
and it could do worse than start with the startling evidence of
individual case histories.*

In the case of neurofeedback, anecdotal evidence points to possible avenues of research in the treatment of alcoholism, stroke, mental retardation, fetal drug abuse, depression, dementia, autism, and head trauma several years post-injury, where spontaneous recovery and significant response to belated treatment has never previously been reported more than two years after injury.

The Association for Applied Psychophysiology and Biofeedback (which includes neurofeedback) today has more than two thousand members and holds well-attended annual meetings, but missing still is consensus of the sort that ought to define even a field in which there are controversies and personality clashes.

Fifty years after Sterman's eureka moment with his cats, neurofeedback remains a discovery that happened too soon, but Sterman himself thinks there is reason for optimism. In an article published in 2013, he and co-author Leslie M. Thompson wrote,

> *Considering the noninvasive and relatively benign nature of EEG
> feedback therapy, as opposed to the common side effects and costs
> associated with lifelong drug dependence, we do not view neu-
> rofeedback treatment as a "last resort" option for drug-resistant
> cases only, but rather as a generally viable alternative consider-
> ation for any patient suffering from [epileptic] seizures.*

The American Academy of Child and Adolescent Psychiatry (AACAP) agrees. The success of treatment of epilepsy by neurofeedback has been so well documented that this group recognizes its use as meeting criteria for evidence-based treatments, which means the AACAP recommends that neurofeedback always be considered as an avenue of treatment. Joel Lubar has published hundreds of studies and succeeded with many patients; among those favorably disposed toward neurofeedback, the Lubar protocol for ADHD is widely regarded as a highly effective neurofeedback treatment for children and adults with ADHD.

Meanwhile, however, while a younger generation of doctors is increasingly willing to suggest non-mainstream measures like neurofeedback to their patients, responsible neurofeedback practitioners are still not thick on the ground. When asked for a recommendation, Sterman finds it possible to name only one or two well-qualified practitioners in most states in the U.S.A.

Could the tide be turning? Could a still-ambiguous situation change with generous funding for effective laboratory testing, experimentation, and widespread, statistical follow-up of cases, as well as greater willingness, among those accustomed to finding their own way, to risk playing by more restrictive rules? Might neurofeedback still transform the practice and success rates of a large and vital field of medicine? Or would more rigorous study show that it cannot, after all, live up to its early promise? The testing that would have told us years ago has not yet happened. Thanks to a disturbingly *un*scientific funding decision, Sterman's science was condemned to disrespect, misuse, and suspicion, without a trial. Othmer remains optimistic:

> *The NIMH stopped funding Sterman's epilepsy research in 1985, arguing that the field had been plumbed. In fact, the field had just begun. One of the most promising findings in this decade of the brain is how amenable the brain is to effecting change in its own function, if only it is given appropriate cues. One of these days all this will be considered obvious. Why shouldn't the brain be able to adapt to new information about itself? It is called learning. That's what our brain does well.*

This optimism may or may not be well placed. At this juncture in the history of medicine, it isn't only a matter of the jury being still out . . . the trial isn't even occurring. And Barry Sterman's breakthrough is not so much lost as just waiting to be found.

NOTES

CHAPTER 1: THE EMPEROR'S NEW ASTRONOMY

3 **His father:** He was the "Shunzhi emperor," the first emperor of the Qing dynasty. He ruled from 1644 to 1661.

4 **dating from 1676:** Noël Golvers and Efthymios Nicolaidis, *Ferdinand Verbiest and Jesuit Science in 17th Century China: An Annotated Edition and Translation of the Constantinople Manuscript (1676)* (Athens, Greece/Leuven, Belgium: Institute for Neohellenic Research/Ferdinand Verbiest Institute, 2009). This book systematically describes Verbiest's Chinese "Instrument-Book."

4 **Wanli Emperor:** He reigned from 1573 to 1620 and was the last emperor of the Ming dynasty.

4 **clocks and harpsichords:** Noël Golvers, *Ferdinand Verbiest, S. J. (1623–1688) and the Chinese Heaven: The Composition of his Astronomical Corpus and its Reception in the European Republic of Letters (Louvain Chinese Studies)* (Belgium: Leuven University Press, 2003), p. 15.

5 **no more than one minute:** Golvers and Nicolaidis, *Ferdinand Verbiest and Jesuit Science*, p. 83.

6 **float in infinite empty space:** Joseph Needham, *Chinese Astronomy and the Jesuit Mission: An Encounter of Cultures* (London, England: The China Society, 1958), pp. 2, 6. This piece appears as #10 in Cyril Birch, ed., China Society Occasional Papers 1–12, and was originally part of *Science and Civilization in China: Volume 3, Mathematics and the Sciences of the Heavens and the Earth* by Needham with the collaboration of Wang Ling.

6 **their discovery in the West:** Ibid., p. 11.

6 **celestial events could be predicted:** Benjamin A. Elman, *On Their Own Terms: Science in China 1550–1900* (Cambridge, MA, and London, England: Harvard University Press, 2008), p. 63.

7 **as early as 25 to 220 CE:** Ibid., p. 63.

7 **what is now northwestern Iran:** Golvers and Nicolaidis, *Ferdinand Verbiest and Jesuit Science*, p. 57.

7 **the calamities it portended:** Needham, *Chinese Astronomy*, p. 81.

8 **heavenly approval for the emperor:** See Elman, *On Their Own Terms*, reference to Thatcher E. Deane, "The Chinese Imperial Astronomical Bureau: Form and Function of the Ming Dynasty 'Qintianjian' from 1365 to 1627," PhD Thesis (Seattle, WA: University of Washington, 1989), pp. 262–269.

8 **empire were in peril:** Elman, *On Their Own Terms*, p. 6.

8 **produced no improvement:** It was a situation not unfamiliar to European astronomers. In the sixteenth century, when the solstices were arriving late by ten days, Pope Gregory XIII established a commission to propose a reform, and Christopher Clavius, a Jesuit, used tables based on Copernican astronomy to devise the new Gregorian calendar. Ten days were removed from the calendar to make the adjustment.

9 **on occasion, even came to services:** Editors of *Encyclopædia Britannica*, online, s.v. "Adam Schall von Bell" (Chicago: Encyclopædia Britannica, 2008); http://www.britannica.com/EBchecked/topic/527026/Adam-Schall-von-Bell?anchor=ref1003343.

10 **"more than I taught":** Golvers, *Ferdinand Verbiest, S. J.*, p. 22. Verbiest to Athanasius Kircher, 1659.

10 **"and the rise of the stars":** Ibid., p. 22. Verbiest to Kircher, 1659.

11 **arrived in the early 1800s:** Needham, *Chinese Astronomy*, p. 11.

13 **"especially in mathematics":** Golvers, *Ferdinand Verbiest, S. J.*, p. 19.

14 **a peaceful old age:** Golvers and Nicolaidis, *Ferdinand Verbiest and Jesuit Science*, pp. 112–163. The information that follows is taken from Verbiest's own account in his *Compendium Historicum de Astronomia apud Sinas restituta*, reprinted and translated as *Historical Compendium on the Restitution of Astronomy in China* in Golvers and Nicolaidis.

15 **his infant son's burial:** Nicolas Standaert, *The Interweaving of Rituals: Funerals in the Cultural Exchange between China and Europe* (Seattle, WA: University of Washington Press, 2008).

15 **neither stand nor sit:** Joseph F. MacDonnell, "Fr. Ferdinand Verbiest, S.J. (1623–1688) a Jesuit Scientist in China," Fairfield University Mathematics Department; http://www.faculty.fairfield. edu/jmac/sj/scientists/verbiest.htm.

16 **"teeming with mistakes":** Golvers and Nicolaidis, *Ferdinand Verbiest and Jesuit Science*, pp. 115ff. Verbiest's own account.

16 **his own and his colleagues' lives:** Ibid., p. 113.

17 **"in the calculations of the calendar":** Ibid., p. 117.

18 **the shadow would be long:** Ibid., pp. 119, 294 (figure). Verbiest's detailed description of the apparatus, with the drawing to illustrate, can be found in his account of the events.

18 **The First Gnomon Experiment:** III. No. 1 of the Metochion manuscript, figure reproduced in Golvers and Nicolaidis, *Ferdinand Verbiest and Jesuit Science*, used with permission of Dr. Noël Golvers.

18 **The Second Gnomon Experiment:** III. No. 2 of the Metochion manuscript, figure reproduced in Golvers and Nicolaidis, *Ferdinand Verbiest and Jesuit Science*, used with permission of Dr. Noël Golvers.

20 **"at each other with pale faces":** Verbiest, in Golvers and Nicolaidis, *Ferdinand Verbiest and Jesuit Science*, p. 125.

21 **"between the shadow and my calculations":** Ibid., pp. 112–163.

21 **"my diligence and assiduity!":** Ibid., p. 127.

22 **"deviated from the Heavens":** Ibid., p. 133 and note 78 on p. 159.

22 **"former accusation of Adam Schall was true":** Willy Vande Walle, "Ferdinand Verbiest and the Chinese Bureaucracy," in *Ferdinand Verbiest, S. J. (1623–1688): Jesuit Missionary, Scientist, Engineer and Diplomat*, ed. John W. Witek, S. J. (Leuven, Belgium: Institut Monumenta Serica, 1994), p. 499.

23 **overpowered him:** Kangxi is the Westernization of the name K'ang-hsi, who was also known as Xuanye, and Shengzu. His posthumous name was Rendi. Editors of *Encyclopædia Britannica*, online, s.v. "Kangxi: Emperor of Qing Dynasty." *Encyclopædia Britannica*. Chicago: Encyclopædia Britannica, Inc., 2008. http://www.britannica.com/biography/Kangxi.

23 **"must necessarily be cancelled":** Verbiest, in Golvers and Nicolaidis, *Ferdinand Verbiest and Jesuit Science*, p. 137.

23 **"in so many matters!":** Ibid., p. 137.

23 **new calendars to the Imperial Palace:** Ibid., pp. 143–147.

26 **beautiful edition of Verbiest's** *Astronomia Europaea:* The full title is *Astronomia Europaea sub Imperatore tartaro-sinico Cam-hy appellato ex jmbra in lucem revocata*. It is available now with an English translation as *The Astronomia Europaea of Ferdinand Verbiest, S. J. (Dillingen, 1687)* in both Latin and English text, with translation, notes, and commentaries. Noel Golvers, *The Astronomia Europaea of Ferdinand Verbiest, S. J. (Dillingen, 1687): Text, Translation, Notes and Commentaries* (Nettetal, Germany: Steyler Verlag, 1993).

27 **causing the wheel to turn:** Verbiest didn't invent the *aeolipyle*, the steam engine, the gearing system he used, or the tiller design. It was his genius to combine them in a completely new and highly creative way for a previously unimagined purpose.

28 **popular in China at the time:** Ibid., p. 259.

28 **his "automotive machine":** Drawing from John W. Witek, ed., *Ferdinand Verbiest, S.J. (1623–1688): Jesuit Missionary, Scientist, Engineer and Diplomat* (Leuven, Belgium: Institut Monumenta Serica, Sankt Augustin and Ferdinand Verbiest Foundation, 1994). Used with permission of the Institut Monumenta Serica.

28 **a room or courtyard:** Ditlev Scheel has gone to great lengths in a project to recreate Verbiest's automobile and find out whether such an apparatus could actually have worked. He has described his project in detail in his article "Beijing Precursor." What sounds relatively simple in Verbiest's account turns out to be undoubtedly workable but not at all easy or uncomplicated to achieve. Scheel's article is in Golvers, *Ferdinand Verbiest, S.J. (Dillingen, 1687)*, pp. 120–124, 430–436, 496–497.

28 **a flight of imagination:** The description of the vehicle's carrying a human passenger is reported in Kenneth Richardson, *The British Motor Industry 1896–1939: A Social & Economic History* (London, England: Macmillan, 1977), p. 1.

29 **"eyes of such a great majesty":** *The Astronomia Europaea of Ferdinand Verbiest, S. J.*, p. 101.

29 **Verbiest's output . . . was prodigious:** Excepting the telescope. A telescope had been brought to China in 1618 by the astronomer Johannes Terrentius (Johann Schreck) and was given to the emperor in 1634, but it seems to have made no significant impression. Needham, *Chinese Astronomy*, p. 8.

29 **elaborately decorated:** Ibid., p. 15.

30 **"bitterness, anger, or recrimination":** Francis A. Rouleau and Edward J. Malatesta, "The 'Excommunication' of Ferdinand Verbiest," in Witek, *Ferdinand Verbiest, S. J.* p. 494. This dispute was a complicated, lengthy series of arguments and correspondence. Rouleau and Malatesta have described it in detail with many relevant footnotes and references. Verbiest was never excommunicated or in danger of it.

30 **"they render[ed] us great prestige and respect":** Golvers and Nicolaidis, *Ferdinand Verbiest and Jesuit Science*, p. 265.

31 **"a protector for the holiest religion in the universe":** H. Josson and L. Willaert, *Correspondance de Ferdinand Verbiest de la Compagnie de Jésus (1623–1688)* (Brussels, 1938), p. 551. Quoted in translation in Witek, *Ferdinand Verbiest, S. J.*, p. 20.

CHAPTER 2: FARM LAD, SPY, ARISTOCRAT, RASCAL

34 **and caricature people:** Marc-Auguste Pictet, a close friend of Thompson's who visited him for several days in London in 1801, took notes on their conversations. Some of Pictet's reporting was inaccurate—for example, the reporting of Thompson's birthplace. George Ellis drew carefully and discriminatingly from Pictet's account when he wrote his book *Memoir of Sir Benjamin Thompson, Count Rumford, with Notices of his Daughter*, published in 1871. Some of Thompson's sketches were preserved in a notebook handed down from the father of Thompson's first wife, Reverend Timothy Walker, which George Ellis had on hand. George E. Ellis, *Memoir of Sir Benjamin Thompson, Count Rumford, with Notices of His Daughter* (Boston: American Academy of Arts and Sciences, 1871).

34 **passion for music:** Ellis, *Memoir of Sir Benjamin Thompson*, p. 17.

35 **"fondness for experimental philosophy":** George Atkinson Ward, *The Journal and Letters of Samuel Curwen, Judge of Admiralty* (New York, NY: C.S. Francis 1842), p. 497.

35 **"good natured and sensible":** Ibid., p. 316.

35 **"always with increased interest":** Sanborn C. Brown, editor, *Count Rumford, The Collected Works of Count Rumford, Volume I: The Nature of Heat* (Cambridge, MA: The Belknap Press of Harvard University Press, 1968), p. 443. Also quoted in Sanborn C. Brown, "Count Rumford and the Caloric Theory of Heat," *Proceedings of the American Philosophical Society*, Volume 93, No. 4 (1949): 324. Published by: American Philosophical Society, quoted from Rumford, *Mémoires sur la chaleurs*, Paris; http://www.jstor.org/stable/3143157.

35 **"and on reasonable terms":** Quoted in Sanborn C. Brown, *Benjamin Thompson, Count Rumford* (Cambridge, MA: MIT Press, 1979), p. 7.

36 **Loammi Baldwin:** Baldwin spied for the American side of the Revolution, which did not prevent the two from remaining fast friends.

36 **"to make himself agreeable":** Ellis, *Memoir of Sir Benjamin*, p. 43.

37 **"all your winter's earnings on finery?":** Ibid., p. 45.

38 **"true interest of this my native Country":** Sanborn C. Brown, *Benjamin Thompson, Count Rumford*, pp. 32–33.

39 **Improvement of Natural Knowledge:** See footnote in Sanborn C. Brown, editor, *Count Rumford, The Collected Works*, p. 444.

40 **"branches of polite learning":** Charles Richard Weld, *A History of the Royal Society: With Memoirs of the Presidents*, Vol. II (Cambridge: Cambridge University Press, 2011), p. 212. Originally published 1848.

41 **"beyond his most sanguine expectations":** Ward, *The Journal and Letters*, p. 316.

41 **this news item reached London newspapers:** Sanborn C. Brown, *Benjamin Thompson, Count Rumford*, p. 87.

43 **"descendants of this outraged community."** Quoted in Ellis, *Memoirs of Sir Benjamin*, p. 138, from Nathaniel S. Prime, *A History of Long Island* (New York, NY: Robert Carter, 1845), pp. 250–251. Sanborn C. Brown, in *Benjamin Thompson, Count Rumford*, p. 88 also cites many other sources that give what seem to be authentic accounts of these activities, and G. I. Brown mentions a report from the American secret agent known as "W." G. I. Brown, *Scientist, Soldier, Statesman, Spy: Count Rumford – The Extraordinary Life of a Scientific Genius* (Phoenix Mill: Sutton Publishers, 1999), p. 35.

43 **"designed for my own investigations":** Sanborn C. Brown, editor, *Count Rumford: The Collected Works*, p. 447.

44 **"subordinates and inferiors":** Ellis, *Memoirs of Sir Benjamin*, p. 162.

44 **"all out upon that occasion":** Sanborn C. Brown, editor, *Count Rumford, The Collected Works, Volume V: Public Institutions* (Cambridge, MA: The Belknap Press of Harvard University Press, 1968), p. 3.

45 **making them happy *first*:** Ellis, *Memoirs of Sir Benjamin*, p. 179.

45 **what he had done for them:** Hugh Rowlinson, "The Contribution of Count Rumford to Domestic Life in Jane Austen's Time," *Jane Austen Society of North America: Persuasions On-Line* 23, No. 1 (Winter 2002); http://www.jasna.org/persuasions/on-line/vol23no1/rowlinson.html.

45 **"the will of a few sound heads":** Sanborn C. Brown, *Benjamin Thompson, Count Rumford*, p. 100. Brown cites G. Cuvier, *Recueil des éloges historiques*, Paris: Didot Fréres, 1861, 252.

46 **experiments in the palace gardens:** Sanborn C. Brown, *Benjamin Thompson, Count Rumford*, p. 116.

47 **hung when not in use:** He also recommended designing the pot's lid so that it could hang on a projection of the same hook.

48 **converted in London alone:** Rowlinson, "The Contribution of Count Rumford."

50 **"was contracted to a Rumford":** Jane Austen, *Northanger Abbey*, reprint edition (London, England: Wordsworth Classics, 1992), pp. 161–162.

50 **manufactured in the United States:** See www.rumford.com.

50 **"I had ever tasted before":** Sanborn C. Brown, editor, *Count Rumford: The Collected Works, Volume III: Devices and Utensils* (Cambridge, MA: The Belknap Press of Harvard University Press, 1969), p. 77.

50 **"something less than one penny farthing":** Count Rumford, letter to Lady Palmerston, November 1794. In the collection of Dartmouth College.

50 **principal however she pleased:** Sanborn C. Brown, *Benjamin Thompson, Count Rumford*, p. 171.

50 "heat or light": Sanborn C. Brown, editor, *Count Rumford: The Collected Works, Volume I*, p. v.

51 "preservation of the Town and the Country": Count Rumford, letter to Lady Palmerston, September 25, 1794. In the collection of Dartmouth College.

51 atop the city walls: Sanborn C. Brown, *Benjamin Thompson, Count Rumford*, p. 181.

52 "most curious operations of Nature": Sanborn C. Brown, editor, *Count Rumford: The Collected Works, Volume I*, p. 3.

53 "it actually boiled!": Ellis, *Memoirs of Sir Benjamin*, p. 480.

53 "among the particles of the body": Quoted from Rumford, *Mémoires sur la Chaleurs* in Sanborn C. Brown, "Count Rumford and the Caloric Theory of Heat," p. 324.

53 "all kinds of motion among them": Ibid.

53 "New Experiments Upon Heat": Colonel Sir Benjamin Thompson, in a letter to Sir Joseph Banks, "New Experiments Upon Heat," *Philosophical Transactions of the Royal Society of London*, Vol. 76 (January 1786): 273–304.

53 "cannot be a *material substance*": Sanborn C. Brown, "Count Rumford and the Caloric Theory of Heat," p. 324.

53 "which is Excited by Friction": Benjamin Count of Rumford, "An Inquiry concerning the Source of the Heat Which is Excited by Friction," *Philosophical Transaction of the Royal Society of London*, Vol. 88 (January 1798): 80–102.

54 "represented by 1034 foot-pounds": James Prescott Joule, "On the Mechanical Equivalent of Heat," *Philosophical Transaction of the Royal Society of London* 140 (January 1850): 61–82.

54 "how little can be known": Sanborn C. Brown, *Benjamin Thompson, Count Rumford*, p. 187, citing Essay VII.

54 "more importance than any I ever penned": Count Rumford, letter to Marc-Auguste Pictet, June 9, 1797. In the collection of Dartmouth College.

55 "arrogant and supercilious": Peter Pindar (John Walcot), *The Works of Peter Pindar, Esq. to which are prefixed Memoirs of the Author's Life* Volume II (London, England: 1802).

57 had won Lavoisier's wife: John H. Lienhard, "Engines of Our Ingenuity No. 4: Count Rumford," *University of Houston*. http://www.uh.edu/engines/epi4.htm.

59 "subject of thermotics": *Thermotics* is a term that was used for the study of heat.

59 "an English monopoly": Ellis, *Memoir of Sir Benjamin*, pp. 485–486, quoting from Edward L. Youmans, *The Correlation and Conservation of Forces* (New York, NY: D. Appleton and Company, 1865. One wonders why a professor in Edinburgh was such a champion of "English" science.

CHAPTER 3: PURSUING VENUS

61 Edmond Halley: Two decades later, he would identify the comet that now bears his name.

63 "gregarious and amiable": Quoted in Mark Anderson, *The Day the World Discovered the Sun* (Boston, MA: Da Capo Press, 2012), and translated by him from "Éloge de M. l'Abbé Chappe" in *Histoire de L'Académie Royale des Sciences* (Paris: 1772), p. 171.

63 perhaps more expendable man: Biographical information is from Andrea Wulf, *Chasing Venus: The Race to Measure the Heavens* (New York, NY: Vintage Books, 2013), citing Jean-Paul Grandjean de Fouchy, "Éloge de M. L'Abbé Chappe d'Auteroche," *Histoire & Mémoires*, 1769, pp. 163ff; and Angus Armitage, "Chappe d'Auteroche: A Pathfinder for Astronomy," *Annals of Science* 10, No. 4 (1954): 278ff.

63 an expedition to Siberia: Wulf, *Chasing Venus*, pp. 42–43, and notes. Also Anderson, *The Day the World*, p. 6.

63 to study the Venus transit: Anderson, *The Day the World*, p. 6.

64 "appointed by the Académie": Chappe d'Auteroche, *A Journey into Siberia, Made by Order of the King of France* (London, England: T. Jefferys, 1770), p. 1; http://hdl.handle.net/2027/uc2.ark:/13960/t1dj5pz7d. Chappe's complete journal is available in facsimile online at http://babel.hathitrust.org/cgi/pt?id=uc2.ark:/13960/t1dj5pz7d;view=1up;seq=29.

64 "large apparatus of instruments": Ibid.

65 "on account of the fogs": Ibid., p. 2.

65 "if a shipwreck should happen": Ibid., p. 3.

66 "not one single inn": Ibid., p. 10.

66 "so very hospitable": Ibid.

66 "handsome and amicable": Ibid.

66 "riding-dress": Ibid.

66 "strictly virtuous": Ibid.

66 "of true principles of religion": Ibid.

66 "exposed to every invader": Ibid., p. 11.

68 "the top of the coach by which we got out": Ibid., p. 24.

68 and bought four sledges: Ibid., p. 25.

69 "protector of the Sciences": Ibid., p. 26.

69 "the protection of this Princess": Ibid.

69 "which are most in common use": Ibid.

69 horses harnessed five abreast: Ibid., p. 28.

70 "at a distance a brick wall": Ibid., p. 35.

70 "to prevent debauchery": Ibid., p. 37.

70 "with inconceivable swiftness": Ibid., p. 38.

70 "in the summer time": Ibid.

70 "motion while I stood upright": Ibid.

71 "royal post": Ibid.

71 "these hamlets": Ibid.

71 "what was their due": Ibid.

71 and riding postilion: Ibid.

71 "chapel": Ibid., p. 44.

72 "longer than they intended": Ibid., p. 47.

72 "concerning these baths": Ibid., p. 54.

72 mercury was not an animal: Ibid., p. 76.

73 "side of the river in an instant": Ibid.

74 "a telescope nineteen feet long": Ibid., p. 79.

74 establish the exact longitude of Tobolsk: A lunar eclipse (unlike a solar eclipse) occurs at the same time everywhere on the face of the earth, making it possible to synchronize time measurements in locations vast distances apart.

75 "one single black cloud": d'Auteroche, *A journey into Siberia*, p. 81.

75 "deprived of by a cloud": Ibid.

75 "serenity of the air": Ibid., p. 83.

75 "in order not to miss it": Ibid.

75 the moment when it last touched the sun's rim: For a technical explanation of the use of a Venus transit to find solar parallax, see F. Mignard, "The Solar parallax with the transit of Venus," Observatoire de la Côte d'Azur, Le Mont Gros BP 4229 06304 Nice Cedex 4, Version 4.1, December 22, 2004; https://www-n.oca.eu/Mignard/Transits/Data/venus_contact.pdf.

76 as soon as it arrived there: Published as *Mémoire du passage de Vénus sur le soleil, avec des observations sur l'astronomie et la déclinaison de la boussole faites à Tobolsk, en Sibérie.*

76 like a drop of water or the tip of a rapier: Wulf, *Chasing Venus*, pp. 92–93.

78 a splendid publication: Published as *Voyage en Sibérie fait en 1761 (avec la description du Kamtschatka, trad. du russe de Khracheninnikow)*, Chappe d'Auteroche, 1768.

79 "cruel and inhuman": d'Auteroche, *A Journey into Siberia*, p. 317.

79 "common people": Ibid., p. 318.

79 "greatest pleasure they have": Ibid.

79 a scathing pamphlet appeared: *Antidote ou Réfutation du mauvais livre superbement imprimé intitule: Voyage en Sibérie, etc.* Probably authored by Count Andrey Petrovich Shuvalov, though rumored to have been written by Catherine the Great.

80 "King's Engineer and Geographer": Chappe d'Auteroche, *A Voyage to California: to Observe the Transit of Venus* (London, England: E. and C. Dilly, 1773), p. 2; http://hdl.handle.net/2027/ nyp.33433000631253. This book is available in facsimile online at http://babel.hathitrust.org/ cgi/pt?id=nyp.33433000631253;view=1up;seq=31.

80 "that was curious": Ibid.

80 "little mischiefs": Ibid.

80 to measure water density: Anderson, *The Day the World*, p. 102, citing Angus Armitage, "Chappe D'Auteroche: A Pathfinder for Astronomy," *Annals of Science* 10, No. 4 (1954): 288–291.

80 "the learned world for this loss": d'Auteroche, *A Voyage to California*, p. 3.

80 "sport of the smallest wave": Ibid., p. 9.

80 "were wanting in the days": Ibid., p. 15.

80 "little nut-shell": Ibid., p. 9.

81 "most admirable and sublime": Ibid., pp. 12–13.

82 "especially for a French-man, to touch a bit": Ibid., p. 24.

82 "disagreeable and unwholesome": Ibid., p. 20.

82 "verdure of rich cultivated land": Ibid., p. 38.

82 "than a religious ceremony": Ibid., p. 39.

83 "whose first precept is Charity": Ibid., p. 45.

84 "retarded us more than once": Ibid., p. 47.

84 "noble," "very solid" aqueduct: Ibid., p. 48.

84 "discern its origin and direction": Ibid., p. 50.

84 "most frightful situation": Ibid., p. 51.

84 "with terror and admiration": Ibid.

84 "fir-trees in the vale below": Ibid., pp. 51–52. Chappe's description will sound familiar to those who, even in the twenty-first century, choose to avoid the motorways in Mexico and travel by the old mountain roads.

85 "waves should chance to wash us": Ibid., p. 61.

85 "at last we got safe to land": Ibid., pp. 61–62.

85 "it is impossible to express": Ibid., p. 62.

86 "an epidemical distemper": Ibid., p. 79.

86 "a great beam of cedar": Ibid., p. 54.

86 mechanism from wind and dust: For Chappe's detailed preparation and his list, see Anderson, *The Day the World*, p. 184.

86 "setting of my clock": d'Auteroche, *A Voyage to California*, p. 63.

86 "Mr. Chappe cared for nothing else": Ibid., p. 65.

87 the transit was over: Details about the activities on the day of the transit come from "On the Observations made by Abbé Chappe in California" and "Astronomical Observations Made in the Village of San José in California," both works transcribed and translated in Doyce B. Nunis, Jr., editor, *The 1769 Transit of Venus: The Baja California Observations of Jean-Baptiste Chappe d'Auteroche, Vicente de Doz, and Joaquín Velázquez Cárdenas de León* (Los Angeles, CA: Natural History Museum of Los Angeles County, 1982).

87 **"most complete observation":** d'Auteroche, *A Voyage to California*, p. 63.

87 **Pauly wrote:** Ibid., p. 66.

87 **accuracy of his pendulum clock:** For Chappe's observations after the transit: Anderson, *The Day the World*, pp. 206–209.

87 **on a ship back to the mainland:** Ibid., p. 210.

88 **died soon after landing:** Doz survived and delivered his observations to the king of Spain in the summer of 1770.

88 **"had not been fatal to him":** Quoted in Nunis Jr., *The 1769 Transit*, p. 87, from "Éloge de M. l'Abbé Chappe," in *Histoire de L'Académie Royale des Sciences* (Paris: 1772), p. 171.

88 **93 million miles:** Also close to Halley's hope of 99.2 percent accuracy.

88 **the planet touched the sun's disk:** We now know that the atmosphere of Venus and the sun's corona cause this blurring.

89 **"the wonders it contains":** d'Auteroche, *A Voyage to California*, pp. 13–14.

CHAPTER 4: LOST IN HER OWN LEGEND

93 **his true story:** see Kitty Ferguson, *The Music of Pythagoras* (New York, NY: Walker Books, 2008).

94 **considered reliable documentation:** W. K. C. Guthrie, *A History of Greek Philosophy, Vol. I: The Earlier Presocratics and the Pythagoreans* (Cambridge, England: Cambridge University Press, 1962), p. 16.

95 **twenty-eight-volume chemical encyclopedia:** *Cheirokmeta* was the title. "Mystical sister" were the words Zosimos used to refer to Theosebeia.

95 **"the hinge of legend and history":** Jacob Bronowski, *The Ascent of Man* (Boston, MA: Little, Brown & Co., 1973), p. 120.

97 **"these things in a secret language":** Quoted in Raphael Patai, *The Jewish Alchemists: A History and Source Book* (Princeton, NJ: Princeton University Press, 1994 and 2014), p. 51, from Zosimos, who was in turn quoting Pseudo-Democritus. Much of the information in this chapter comes through Zosimos, as translated and interpreted by Patai.

97 **"anything clearly in this regard":** Ibid., footnote 17, quoted from Zosimos, *The First Book of the Final Reckoning*.

97 **"the race of Abraham":** Ibid., p. 50, quoting Zosimos.

98 **Egypt during her time:** See discussion of the Mareotic group, a scholarly Jewish community inclusive of both sexes near Alexandria in northern Egypt in the first century AD, in Joan E. Taylor, *Jewish Women Philosophers of First-Century Alexandria: Philo's "Therapeutae" Reconsidered* (Oxford, England: Oxford University Press, 2003).

98 **"the ancients," "the sages," or the "first of the ancient authors":** Patai, *The Jewish Alchemists*, p. 12.

98 **two generations earlier than he did:** See Ibid., p. 60.

100 **Egypt was at their doorstep:** For these two paragraphs about the Jews in Alexandria, I have drawn information from Emil Schürer, "Alexandria, Egypt—Ancient," *Jewish Encyclopedia: The unedited full-text of the 1906 Jewish Encyclopedia*; http://www.jewishencyclopedia.com/articles/1171-alexandria-egypt-ancient.

100 **"ally of the Roman aristocracy":** Raymond F. Surburg, *Introduction to the Intertestamental Period* (St. Louis, MO: Concordia Publishing House, 1975), pp. 155–156.

100 **"strange fusion":** Ibid., p. 157.

100 **"the science of chemistry takes its name":** "Zosimos of Panopolis," *Wikipedia: The Free Encyclopedia*; https://en.wikipedia.org/wiki/Zosimos_of_Panopolis, quoted in William Drummond, "On the Science of the Egyptians and Chaldeans," *The Classical Journal*, Vol. 18 (London, England: A. J. Valpy, 1818), p. 299.

101 "expert in the work": Patai, *The Jewish Alchemists*, p. 453.

101 "see without touching": Ibid., p. 61.

101 "most injurious of metals": Ibid.

101 "sulphurous": Ibid., p. 64.

101 "clay of the philosophers": See Patai, *The Jewish Alchemists*, p. 61. From this clay sealant comes the modern term "hermetically sealed" because in the middle ages Mary was believed to have been the "successor" to the mythological sorcerer Hermes Trismegistus.

102 "pastry cook's proper frying-pan": M. Berthelot and Ch. Em. Ruelle, *Collection des Anciens Alchimistes Grecs* (Paris: Georges Steinheil, 1888).

102 a paste of flour to seal the joints: Quoted in Margaret Alic, *Hypatia's Heritage: A History of Women in Science from Antiquity to the Late Nineteenth Century* (London, England: Women's Press Limited, 1986), p. 37; citing F. Sherwood Taylor, "The Evolution of the Still," *Annals of Science* 5 (1945): 190.

103 attar of roses and other plant oils: Alic, *Hypatia's Heritage*, p. 39.

104 "the dog with the wolf": Ibid., quoting from Zosimos.

105 continues page after page: Patai, *The Jewish Alchemists*, pp. 64–89, quoting Zosimos, who in turn is quoting Mary the Jewess.

105 "also makes it approximate gold": Ibid., p. 86 (see also footnote 18), quoting from Zosimos.

106 "a permanent black tint": Ibid., p. 82, quoting from Zosimos, who is quoting from Mary the Jewess.

106 "the precepts of the Hebrews": Ibid., pp. 56–57, regarding Zosimos's paraphrase of Mary the Jewess in his *The Coloring of Precious Stones*.

106 "One is All and All is One": Alic, *Hypatia's Heritage*, p. 39, quoting from Zosimos, who is quoting from Mary the Jewess.

107 not at the top of the table: Walter Burkert, *Lore and Science in Ancient Pythagoreanism*, (Cambridge, MA: Harvard University Press, 1972), p. 51.

107 and without an opposite: Ibid., p. 54, footnote 7.

107 "God" in the singular: Patai, *The Jewish Alchemists*, p. 70.

107 a very cosmopolitan city: We also find Mary referring to some of the resulting mixtures in her experiments as the "All." What the connection is here is unknown.

107 "thus two are but one": Patai, *The Jewish Alchemists*, p. 66.

108 numbers for unlocking its secrets: Ferguson, *The Music of Pythagoras*, p. 130.

109 also made of four triangles: Ibid., p. 107.

109 (the cube of three): Ibid., pp. 130–131.

110 impoverished his sister, Sophie Brahe: Kitty Ferguson, *Tycho & Kepler: The Unlikely Partnership That Forever Changed Our Understanding of the Heavens* (New York, NY: Walker & Company, 2002), pp. 172–174.

111 particular link with Christianity: See Patai, *The Jewish Alchemists*, p. 75.

111 "Maria the Copt": Alic, *Hypatia's Heritage*, p. 194 mentions in a footnote that Mary the Jewess and Mary the Copt may have been two different people.

111 "daughter of the king of Saba": Patai, *The Jewish Alchemists*, p. 76.

111 "it be forgotten": Ibid.

112 before the years she actually lived: The information and quotations in the paragraph are from Patai, *The Jewish Alchemists*, pp. 75–76.

112 "technical aid" when working with mystical formulas: Robert Lindel, "Music and Patronage at the Court of Rudolf II," in *Music in the German Renaissance: Sources, Styles, and Contexts*, edited by John Kmetz (Cambridge, England: Cambridge University Press, 1994), p. 268.

112 "successor" to Hermes Trismegistus: Patai, *The Jewish Alchemists*, p. 245. Hermes Trismegistus was a mythical Egyptian alchemist.

CHAPTER 5: WONDROUS TRANSFORMATIONS

116 "on which they were found": Maria Sibylla Merian, *Metamorphosis insectorum Surinamensium, in Maria Sibylla Merian in Surinam, Kommentar zur Faksimile-Ausgabe (Commentary to the facsimile edition of Metamorphosis insectorum Surinamensium)*, translated by Elisabeth Rücker and William T. Stearn (London, England: Pion, 1982), p. 120. Based on original watercolors published in Amsterdam in 1705, which are kept in the Royal Library, Windsor.

117 "the heady brew was ink": Kim Todd, *Chrysalis: Maria Sibylla Merian and the Secrets of Metamorphosis* (New York, NY: Harcourt, 2007), p. 20.

117 naturalistic before Maria's: Summary published to introduce the exhibition *Maria Sibylla Merian and Daughters: Women of Art and Science*, June 10–August 31, 2008, at the J. Paul Getty Museum, Los Angeles, CA.

119 "alive through the other": Quoted in Todd, *Chrysalis*, p. 35.

119 "in 1660. Thank God": Maria Sibylla Merian, *Studienbuch*. Now in the St. Petersburg, Russia, Academy of Sciences.

119 "describe them from life": Rücker and Stearn translation, *Maria Sibylla Merian*, p. 85.

119 few believed that all did: The experiments of Franceso Redi were one of the first challenges to spontaneous generation, but Merian's observation and depiction of metamorphosis predated by ten years his *Esperienze Intorno alla Generazione degl'Insetti (Experiments on the Generation of Insects)*, which is often erroneously cited as the first report of it. See summary published to introduce the exhibition *Maria Sibylla Merian and Daughters: Women of Art and Science*.

120 appeared between 1662 and 1667: Johannes Goedaert's *Metamorphosis naturalis* (c. 1662–1667); dealt with the metamorphosis of about 140 insects but did not portray their food plants.

121 "watery material comes flowing out": Quoted in Todd, *Chrysalis*, p. 60.

121 "emerge from her hands daily": Ibid., pp. 66–67, from Joachim von Sandrart, *Teutsche Academie*, 1675.

121 pictures in a tulip catalogue: See ibid., p. 68. Todd has described the works in this book and noted the different styles.

122 under the title *Neues Blumenbuch*: Maria Sibylla Merian, first volume of *Florum fasciculus primus* (1675).

122 "worms, flies, gnats, spiders": Todd, *Chrysalis*, p. 66, from Sandrart.

122 a stroll with a friend: Maria Sibylla Merian, *Der Raupen wunderbare Vervandelung und sonderbare Blumennahrung*, 1769 (her caterpillar book).

122 "on the white dead-nettle": Quoted in Todd, *Chrysalis*, p. 70, from Maria Sibylla Merian, *Der Raupen: Vol. II*.

123 "a new invention": See Susan Owens, "Maria Sibylla Merian: 'Great Diligence, Grace and Spirit,'" in David Attenborough, Susan Owens, Martin Clayton, and Ria Alexandratos, *Amazing Rare Things: The Art of Natural History in the Age of Discovery* (New Haven, CT, and London, England: Yale University Press, 2007), p. 141.

123 *of Matthaus Merian the Elder*: The full title of Maria Sibylla Merian, *Der Raupen*.

123 "world in a little book": Quoted in Natalie Zemon Davis, *Women on the Margins: Three Seventeenth-Century Lives* (Cambridge, MA: Harvard University Press, 1995), p. 156, from the preface to Maria Sibylla Merian, *Der Raupen wunderbare Verwandlung und sonderbare Blumennahrung: Vol. I* (1769).

124 "to the gentlemen scholars": Quoted in Todd, *Chrysalis*, p. 77, from Maria Sibylla Marien, *Der Raupen: Vol. I*.

126 "try to hear or see him": Quoted in Davis, *Women On the Margins*, from Jean de Labadie, *Les Entretiens d'Esprit du Jour Chretien, ou les Reflexions Impotantes du Fidele* (Amsterdam: Laurans Autein, 1671), p. 4.

126 "in heath and moorland": Rücker and Stearn translation, *Maria Sibylla Merian*, p. 85.

128 "Levinus Vincent and many others": Ibid.

128 "and subsequent development": Ibid.

128 "pursue my investigations further": Ibid.

130 "my Indian": Todd, *Chrysalis*, p. 161.

130 at the time of Merian's visit: Ibid., p. 158.

130 preferred their sweaty beds: Todd has given a lengthy, interesting description of the town, houses and lifestyle. Todd, *Chrysalis*, pp. 156ff.

131 "people there die of the heat": Maria Sybilla Merian, letter to Mr. Johann Georg Volkammer, Doctor of Medicine in Nuremburg, October 8, 1702, translated in Maria Sibylla Merian, *Metamorphosis insectorum Surinamensium*, translation of the text of the first edition. In Rücker and Stearn translation, p. 65.

131 vellum pages of her *Studienbuch*: This book is now in the Academy of Sciences in St. Petersburg, Russia.

131 "rendered with the paint brush": Merian, quoted and translated in Owens, "Maria Sibylla Merian," p. 148.

132 "tightly to the trees": Todd, *Chrysalis*, p. 157, quoted from Wolf-Dietrich Beer, editor, *Butterflies, beetles and other insects: the Leningrad book of notes and studies*, (a translation of Merian's *Studienbuch*) (New York, NY: McGraw-Hill International Book Company, 1976), entry 232.

132 "beautiful flies": Ibid.

133 "fiery flame came out": Owens, "Maria Sibylla Merian," p. 159.

133 double-blossomed pomegranate: Figure 8 in David Brafman and Stephanie Schrader, *Insects and Flowers: The Art of Maria Sibylla Merian* (Los Angeles, CA: J. Paul Getty Museum, 2008). Plate 49 in Maria Sibylla Merian, *Metamorphosis insectorum Surinamensium*.

133 tobacco hawk moth and its larva: Figure 3 in Brafman and Schrader, *Insects and Flowers*. Plate 45 in Maria Sibylla Merian, *Metamorphosis insectorum Surinamensium*.

135 during their fierce combat: This painting can be seen in color as the ninth plate in an online slideshow at the website of the exhibit *Maria Sibylla Merian and Daughters: Women of Art and Science*, June 10–August 31, 2008, at the J. Paul Getty Museum in Los Angeles, CA, located at http://www.getty.edu/art/exhibitions/merian/.

135 "with my life": Maria Sibylla Merian, letter to Johann Georg Volkammer, October 1702, in Rücker and Stearn translation.

136 "the plants and creatures life size": Ibid.

136 forty-five florins apiece: Londa Schiebinger, *The Mind Has No Sex?: Women in the Origins of Modern Science* (Boston, MA, and London, England: Harvard University Press, 1989), p. 76.

137 "as a Dutch rusk [Zwieback]": Merian 1705 (Rücker and Stearn translation, p. 91).

137 "investigators of nature": Merian 1705 (Rücker and Stearn translation, p. 84).

137 "who created such wonders": Quoted by Schiebinger, *The Mind*, p. 72, from Merian, citing the introduction to *Metamorphosis*, but I have not found the quotation there.

137 "makes living extremely difficult": Merian, letter to Volkammer, October 1702 in Rücker and Stearn translation.

138 "according to his own judgement": Merian 1705 (Rücker and Stearn translation, p. 85).

138 "with pleasure and joy": Todd, *Chrysalis*, p. 212.

139 "delivered on the third payment": *Philosophical Transactions of the Royal Society*, volume 23 (1702–1703), appearing at the end of number 285, 1703, np.

140 "a very courteous woman": Quoted in Davis, *Women on the Margins*, p. 198.

140 "paint could achieve as much": Quoted in Ibid., p. 182.

141 greater circulation than the originals: Todd, *Chrysalis*, p. 235.

142 his tarantula fled in panic: Ibid., p. 242.

142 only of its "showy" format: Schiebinger, *The Mind*, p. 79.

142 "areas beyond her ken": Todd, *Chrysalis*, p. 245.

143 his translation of *Der Raupen*: Maria Sybilla Merian, *The Wondrous Transformation of Caterpillars: Fifty Engravings Selected from Erucarum Ortus (1718)*, with an introduction by William T. Stern (London, England: Scolar Press, 1978), p. 10.

CHAPTER 6: THE OTHER DARWIN

145 "higher groups of animals?": Alfred Russel Wallace, *My Life: A Record of Events and Opinions*, Vol. 2 (New York, NY: Dodd, Mead & Company, 1905 or 1908), p. 361.

146 "new species would be explained": Ibid., pp. 361–362.

146 his theory in detail: Wallace wrote in his autobiography that he was on the island of Ternate when this happened, but the journal he was keeping at the time indicated he was on Gilolo, an island where he stayed for a month during the time he was using Ternate as his base camp.

146 "Permanent Distinctness of Species": Alfred Russel Wallace, "On the Tendency of Varieties to Depart Indefinitely from the Original Type; Instability of Varieties Supposed to Prove the Permanent Distinctness of Species," *Journal of the Linnean Society (Zoology)* 3 (1858): 53–62. The essay is available online at the Darwin Correspondence Project, "Alfred Russel Wallace's essay on varieties," *University of Cambridge*; https://www.darwinproject.ac.uk/people/about-darwin/origin-species/alfred-russel-wallace-s-essay-varieties. The date given there for the journal article is 1859.

147 "a better short abstract": Charles Darwin, letter to Charles Lyell, June 18, 1858, record number WCP5647, "Wallace Letters Online," *Natural History Museum*; http://www.nhm.ac.uk/research-curation/scientific-resources/collections/library-collections/wallace-letters-online/5647/6498/details.html, record created August 18, 2014.

151 "to oppress the earth": Quoted in Ross A. Slotten, *The Heretic in Darwin's Court: The Life of Alfred Russel Wallace* (New York, NY: Columbia University Press, 2004), p. 49, from Alfred Russel Wallace, "Equatorial Vegetation," in *"Tropical Nature" and Other Essays, 1878*, reprint (New York, NY: AMS Press, 1975), pp. 29–34.

152 restricted and unexplainable: Information about Wallace's time in South America comes primarily (unless otherwise cited) from Alfred Russel Wallace, *A Narrative of Travels on the Amazon and Rio Negro, With an Account of the Native Tribes, and Observations on the Climate, Geology, and Natural History of the Amazon Valley*, 2nd edition (London, England; New York, NY; and Melbourne, Australia: Ward, Lock and Co., Minerva Library of Famous Books, 1889). Facsimile online at https://archive.org/details/travelsonamazonr00wall.

152 publication in a scientific journal: Alfred Russel Wallace, "On the Umbrella Bird" (Cephalopterus ornatus)," *Annals and Magazine of Natural History* Vol. 8 (1851): 428–430, DOI: 10.1080/03745486109494996. Published online December 23, 2009.

152 even live animals: The list in Slotten, *The Heretic*, p. 83, comes from Elbert Hubbard, *Little Journeys to Homes of Great Scientists: Wallace* (New York, NY: Roycrofters, 1905), p. 96.

153 "from those wild regions . . .": Wallace, *My Life*, Vol. 1, p. 307.

154 "rarely absent from my thoughts": Wallace, *My Life*, Vol. 1, p. 354.

154 "closely allied species": Alfred Russel Wallace, "On the Law which has Regulated the Introduction of New Species," *Annals and Magazine of Natural History* 16 (September 1855): 184–196. Available on the Electronic Scholarly Publishing Project at http://www.esp.org/books/wallace/law.pdf.

155 "the how was still a secret": Wallace, *My Life*, Vol. 1, p. 355.

156 "becomes apparent to us": Wallace, "On the Law."

157 "modified prototypes only remain": Ibid.

NOTES

157 "arguments against its probability": Ibid.

157 "even elicited opposition": Alfred Russel Wallace, letter to Charles Darwin, September 27, 1857, record number WCP4080, "Wallace Letters Online," *Natural History Museum*; http://www.nhm.ac.uk/research-curation/scientific-resources/collections/library-collections/wallace-letters-online/4080/4027/details.html, record created March 8, 2012.

157 in a forthcoming book: Ibid.

158 "small fragment of your data": This exchange of letters between Lyell and Darwin is quoted in Slotten, *The Heretic*, p. 122, citing *Sir Charles Lyell's Scientific Journals*, pp. xivii and xivi.

158 "doctrines before me": Ibid.

158 "the base of the bill and the feet": Alfred Russel Wallace, "On the Entomology of the Aru Islands," *Zoologist* 16 (1858): 5889–5894.

158 "great perseverance and energy": Quoted in Slotten from the minutes of the February 23, 1858 meeting of the Zoological Society of London, *Zoologist* 16 (1858): 6040–6042.

159 "offer to send it to any journal": Charles Darwin, letter to Charles Lyell, June 18, 1858.

160 "anything which I wrote him": Charles Darwin, letter to Charles Lyell, June 25, 1858, record number WCP5648, "Wallace Letters Online," *Natural History Museum*; http://www.nhm.ac.uk/research-curation/scientific-resources/collections/library-collections/wallace-letters-online/5648/6499/details.html, record created August 18, 2014.

160 He suggested that Hooker and Lyell present the Society: Charles Darwin, Lyell, Hooker correspondence, June 25, 26, 29, 1858.

161 "laid before the Linnean Society": Quoted in Slotten, *The Heretics*, p. 156, from De Beer, *Charles Darwin and Alfred Russel Wallace*, pp. 91–254, 57–58, 264–267.

161 "of science on which they bear": Thomas Bell, "Presidential Address to the Members of the Linnean Society," *Journal of the Linnean Society (Zoology)* 4 (1859): viii–ix.

161 "for having led to this": Charles Darwin, letter to Charles Lyell, July 18, 1858, record number WCP5651, "Wallace Letters Online," *Natural History Museum*; http://www.nhm.ac.uk/research-curation/scientific-resources/collections/library-collections/wallace-letters-online/5651/6502/details.html, record created August 18, 2014.

162 "a few hours later": Alfred Russel Wallace, letter to Joseph Dalton Hooker, October 6, 1858, record number WCP1454, "Wallace Letters Online," *Natural History Museum*; http://www.nhm.ac.uk/research-curation/scientific-resources/collections/library-collections/wallace-letters-online/1454/4022/details.html, record created March 7, 2012.

162 "eminent men on my return home": Alfred Russel Wallace, letter to Mary Anne Wallace, October 6, 1858, record number WCP369, "Wallace Letters Online," *Natural History Museum*; http://www.nhm.ac.uk/research-curation/scientific-resources/collections/library-collections/wallace-letters-online/369/5914/details.html, record created June 3, 2013.

162 "friends who may be interested": Wallace to Stevens, October 29, 1858, [WCP1705]; See ADD 7339.23, Cambridge University Library, Newton Catalogue online at http://www.lib.cam.ac.uk/newton/vger7/. The excerpt on the "Wallace Letters Online," *Natural History Museum* website does not include this part of the letter.

163 "I am a *little* proud of": Alfred Russel Wallace, letter to George Silk, November 30, 1858, record number WCP370, "Wallace Letters Online," *Natural History Museum*; http://www.nhm.ac.uk/research-curation/scientific-resources/collections/library-collections/wallace-letters-online/370/370/details.html, record created June 1, 2002.

163 "what your impression would be": Charles Darwin, letter to Alfred Russel Wallace, January 25, 1859, record number WCP1841, "Wallace Letters Online," *Natural History Museum*; http://www.nhm.ac.uk/research-curation/scientific-resources/collections/library-collections/wallace-letters-online/1841/5924/details.html, record created June 7, 2013.

165 **In early April 1859:** Charles Darwin, letter to Alfred Russel Wallace, April 6, 1859, record number WCP1842, "Wallace Letters Online," *Natural History Museum;* http://www.nhm.ac.uk/research-curation/scientific-resources/collections/library-collections/wallace-letters-online/1842/5925/details.html, record created June 7, 2013.

165 **"The Ornithology of Northern Celebes":** This piece is online at Wiley Online Library, DOI: 10.1111/j.1474-919X.1860.tb06361.x.

166 **"Hooker, Lyell, Asa Gray, etc.":** Charles Darwin, letter to Alfred Russel Wallace, November 13, 1859, record number WCP1844, "Wallace Letters Online," *Natural History Museum;* http://www.nhm.ac.uk/research-curation/scientific-resources/collections/library-collections/wallace-letters-online/1844/5927/details.html, record created June 7, 2013.

166 **"*Principia* of Newton":** Alfred Russel Wallace, letter to George Silk, September 1, 1860, record number WCP363, "Wallace Letters Online," *Natural History Museum;* http://www.nhm.ac.uk/research-curation/scientific-resources/collections/library-collections/wallace-letters-online/363/363/details.html, record created June 1, 2002.

166 **admirable tone and spirit:** Alfred Russel Wallace, letter to Henry Bates, December 24, 1860, record number WCP374, "Wallace Letters Online," *Natural History Museum*; http://www.nhm.ac.uk/research-curation/scientific-resources/collections/library-collections/wallace-letters-online/374/5916/details.html, record created June 3, 2013.

166 **"better, than I have done it":** Charles Darwin, letter to Alfred Russel Wallace, May 18, 1860, record number WCP1846, "Wallace Letters Online," *Natural History Museum*; http://www.nhm.ac.uk/research-curation/scientific-resources/collections/library-collections/wallace-letters-online/1846/5929/details.html, record created June 7, 2013.

166 **many of these previously unknown:** Alfred Russel Wallace, *The Malay Archipelago*, reprint edition (New York, NY: Oxford University Press, 1986), p. viii.

167 **"trekking across the South Pole":** Slotten, *The Heretic,* p. 186.

170 **"yearning for abstract truth":** Alfred Russel Wallace, "The Limits of Natural Selection as Applied to Man," in *Contributions to the Theory of Natural Selection: A Series of Essays* (London, England: Macmillan, 1870), p. 358.

170 **"survival of the fittest":** Ibid., p. 357.

170 **"grievously":** Ibid., p. 360.

170 **"disadvantage":** Ibid.

170 **"life and organization":** Ibid.

171 **"led to such brilliant results":** Wallace, *Contributions,* p. iv.

173 **"got rid of in any other way":** Alfred Russel Wallace, *On Miracles and Modern Spiritualism: Three Essays* (London, England: James Burns, 1875), pp. vi-vii.

174 ***"Wallace's Theory of Evolution":*** John Langdon Brooks, *Just Before the Origin: Alfred Russel Wallace's Theory of Evolution* (New York, NY: Columbia University Press, 1984).

174 **before reading Wallace's essays:** Ibid., p. 241.

174 ***and Alfred Russel Wallace:*** Arnold Brackman, *A Delicate Arrangement: The Strange Case of Charles Darwin and Alfred Russel Wallace* (New York, NY: Times Books, 1980).

175 **"setting matters straight":** Alfred Russel Wallace, acceptance speech, printed in *The Darwin–Wallace Celebration,* Thursday, July 1, 1908 (London, England: Printed for the Linnean Society, Burlington House, Piccadilly, W, 1908).

175 **"whatever in the discovery":** Ibid.

CHAPTER 7: ESCAPE TO OBSCURITY

177 **"not beautiful":** Laurence Badash, ed., *Rutherford and Boltwood: Letters on Radioactivity* (New Haven, Yale University Press, 1969), p. 206.

178 "accompaniment to my life": Lise Meitner, "Looking Back," *Bulletin of the Atomic Scientists* 20 (November 1964): 2.

179 judgeships and the civil service: Ruth Lewin Sime, *Lise Meitner: A Life in Physics* (Berkeley, CA, and London, England: University of California Press, 1996), p. 6.

179 had all their children baptized: At age thirty, Meitner converted to Protestantism.

179 "but think for yourself!": Quoted in Patricia Rife, *Lise Meitner and the Dawn of the Nuclear Age* (Boston, MA: Birkhäuser, 1999), p. 2, citing Charlotte Kerner, *Lise, Atomphysikerin: Die Lebensgeschichte der Lise Meitner* (Wernheim and Basel: Beltz and Gelberg, 1987).

179 "a real course of study": Rife, *Lise Meitner*, p. 5.

180 "through the room without studying!": Lise Meitner, letter to Fräulein Hitzenberger, April 10 and 29, 1951, *Meitner Collection Churchill College Archives Center*, Cambridge, England, MTNR 5/8. Translation by author. Also, Otto Frisch, *What Little I Remember* (Cambridge, UK: Cambridge University Press, 1979), p. 3.

180 "will get out alive": Meitner, "Looking Back," p. 3.

181 no danger of dozing off: Ibid.

181 "the most you have the power to give, yourself": Engelbert Broda, *Ludwig Boltzmann: Mensch, Physiker, Philosoph* (Wien: Franz Deuticke, 1986), p. 21.

181 "a vision she never lost": "Lise Meitner, A physicist who never lost her humanity," Powerpoint presentation, *Department of Physics & Astronomy, The University of Utah*; http://www.physics. utah.edu/~jui/3375/Class%20Materials%20Files/y2007m09d24/LiseMeitner.pdf. Quoting from Frisch, *What Little I Remember*.

181 "'Ave y'seen one of 'em?": Broda, *Ludwig Boltzmann*, p. 97.

182 hadn't yet occurred to her: Meitner, "Looking Back," p. 3.

183 "could be said of Planck": Ibid.

185 "politically too carefree": Quoted and translated in Sime, *Lise Meitner*, p. 35, citing Lise Meitner, "Einige Erinnerungen an das Kaiser-Wilhelm-Institut für Chemie in Berlin-Dahlen," *Naturwissenschaften* 41 (1954): pp. 970–999.

185 "welcomed the other's success": Meitner, "Looking Back," p. 5.

185 "in the way of knowledge and learning": Ibid.

185 "I thought you were a man!": Ibid.

186 "charm of the lady": Badash, p. 206.

187 "wave and corpuscular theories": Albert Einstein, "The Development of Our Views on the Nature and Constitution of Radiation" (Salzburg, Austria: 1909).

187 "I remember the lecture very well": Meitner, "Looking Back," p. 4.

188 "negative thoughts": Quoted in Sime, *Lise Meitner*, from a letter from Meitner to Jenny (last name unknown), February 6, 1911, *Meitner Collection at Churchill College*, Cambridge, England. Reproduced in Jost Lemmerich, ed., *Die Geschichte der Entdeckung der Kernspaltung: Ausstellungskatalog* (Berlin: Technische Universität Berlin, Universitätsbibliothek, 1988), p. 58.

188 "the worst loneliness of all": Ibid.

188 "against academic women": Quoted in Rife, *Lise Meitner*, p. 46, from Lise Meitner, "Status of Women in the Professions," *Physics Today* (August 1960): 20.

188 what they were smelling: Arnold Kramish, *The Griffin: The Greatest Untold Spy Story of World War II* (Boston, MA: Houghton Mifflin Company, 1986), p. 48.

189 radioactive contamination: Previously Meitner and Hahn had handled their substances with bare hands, and samples had regularly been sent back and forth in the mail in cardboard boxes between them and Rutherford.

190 "one of the greatest discoveries!": Georg von Hevesy, letter to Ernest Rutherford, October 14, 1913, Rutherford Papers, Cambridge University, Cambridge, England.

190 (as Bohr himself called it): Rife, *Lise Meitner*, p. 55.

190 "it is derived": Quoted in Sime, *Lise Meitner*, p. 49, citing Otto Hahn and Lise Meitner, "Die Muttersubstanz des Actiniums," *Physikalische Zeitschrift* 19 (1918): 208–218.

191 "about my patients": Rife, *Lise Meitner*, p. 62, quoting Meitner in a letter to Otto Hahn, October 14, 1915, citing the literary estate of Otto Hahn, *Max-Planck-Gesellschaft Archives*, Berlin-Dahlem, Germany.

191 chemical elements: Meitner, "Looking Back," p. 7.

193 "energy of the atom [are] fruitless": Quoted in Rife, *Lise Meitner*, p. 104, citing *The Pittsburgh Post Gazette*, December 9, 1934.

193 "is talking moonshine": Quoted in Rife, *Lise Meitner*, p. 104, from a quotation in Eugene Hecht, *Physics in Perspective* (New York, NY: Addison-Wesley, 1962).

194 "release of nuclear reactions": Quoted in Rife, *Lise Meitner*, p. 141, citing Lise Meitner, "Wege und Irrwege zur Kernenergie," *Naturwissenschaften Rund* 16 (May 1963): 167. Published in English as "Right and Wrong Roads to the Discovery of Nuclear Energy," *Advancement of Science* 19 (1963).

194 "very complicated indeed": Otto Hahn, *Mein Leben*, translated by E. Kaiser and E. Wilkins as *Otto Hahn: My Life* (London, England: MacDonald & Co., 1970), p. 148.

194 "The Jewess endangers the Institute": "Lise Meitner, A physicist who never lost her humanity."

195 "thrown me out": Meitner's diary, March 20–21, 1938, Meitner Collection, Churchill College Archives Center, Cambridge, UK. MTNR 2/15. Translation from German by the author.

195 "the famous scientist Lise Meitner": Rife, *Lise Meitner*, p. 165.

196 "against the interests of Germany": Ibid., p. 166.

197 "friendly persuasion": Ibid.

197 "put everything possible into my trunks": Lise Meitner, letter to Paul Rosbaud, August 4, 1946, *Meitner Collection at Churchill College Archives Center*, Cambridge, England. MTNR 5/15, Document No. 44. Translation from German by the author.

197 "and brought back": Hahn, *Mein Leben*, p. 149.

198 Coster's wife: Information about Meitner's escape and Rosbaud's role in it: Kramish, *The Griffin*, p. 49.

198 "inwardly torn apart": Fokker to Miep Coster, July 16, 1938, Adriaan Fokker Papers, Museum Boerhaave, Leiden. In Sime, *Lise Meitner*, p. 205.

198 "the abduction of Lise Meitner": Wolfgang Pauli, telegram to Dirk Coster, July 1938.

199 "but it is real": Lise Meitner, letter to Otto Hahn, August 24, 1938, *Churchill College Archives Center*, Cambridge, England. MTNR 5/21A. Translation from German by the author.

199 produce radium isotopes: An *isotope* is an atom that has more or fewer neutrons (meaning that it has a greater or lesser atomic mass) than the usual number for that element.

199 "contaminations are playing a joke on us": Otto Hahn, letter to Lise Meitner, December 19, 1938, "Meitner Collection," *Churchill College Archives Center*, Cambridge, England. MTNR 5/21A. Translation from German by the author.

200 "might elongate and divide itself": Otto Frisch, "The Discovery of Fission: How It All Began," *Physics Today* 20, No. 11 (November 1967): 47.

200 "into two smaller ones": In 1937, Niels Bohr had "described the atomic nucleus as a liquid drop with surface tension keeping it together and giving it shape." In "Lise Meitner, A physicist who never lost her humanity."

200 "effect of the surface tension": Frisch, *What Little I Remember*, p. 116.

200 with sticks in the snow: From a conversation with my neighbor in Cambridge, June Johnson, who knew Frisch at the Cavendish Labs.

201 "mass defect in such a break-up": Frisch, "The Discovery of Fission", p. 47.

201 **"it all fitted!":** Otto Frisch, *What Little I Remember*, p. 116.

201 **"provided the correct interpretation":** Albert Einstein, *Out of My Later Years* (New York, NY: Philosophical Library, 1950).

202 **appeared for the first time:** Lise Meitner and Otto R. Frisch, "Disintegration of Uranium by Neutrons: A New Type of Nuclear Reaction," *Nature* 143 (February 11, 1939): 239–240.

203 **"the earlier uranium work":** Lise Meitner, letter to Otto Hahn, March 10, 1939, "Meitner Collection," *Churchill College Archives Center*, Cambridge, England. MTNR 5/21B. Translation from German by the author.

204 **"nothing to do with a bomb!":** Quotation and other information in this paragraph come from "Lise Meitner, A physicist who never lost her humanity."

204 **"war technicians have put our discoveries":** Quoted in "Lise Meitner, A physicist who never lost her humanity."

206 **"'long-term coworker' of mine?":** Quoted in Rife, *Lise Meitner,* pp. 264–265, citing a speech in which it was quoted, delivered in January 1988 by Fritz Krafft at the Hahn-Meitner Institute, Berlin.

207 **and animal protein:** Kramish, *The Griffin*, p. 187.

CHAPTER 8: NEAR-FATAL FICTION

212 **in Plutarch's story:** However, in addition to the story, Plutarch's book *The Face of the Moon* included what amounted to a symposium of ancient scientific thought, presenting the views of Hipparchus, Aristotle, and Aristarchus of Samos.

212 **"advantages which Copernicus has over Ptolemy":** Introduction to Johannes Kepler, *Mysterium*. Quoted in Kitty Ferguson, *Tycho and Kepler*, (New York, NY: Walker & Company, 2003), p. 158, from 1:10 in Max Caspar, Walther von Dyck, Franz Hammer, and Volker Bialas, eds., *Johannes Kepler Gesammelte Werke*, 22 vols. (Munich: Deutsche Forschungsgemeinschaft, and the Bavarian Academy of Sciences, 1937–). Hereafter referred to as JKGW.

213 **"can be expected from him":** JKGW, 13:4.

214 **precedent for his rejected disputation:** See Gale E. Christianson, "Kepler's *Somnium*: Science Fiction and the Renaissance Scientist," Science Fiction Studies, DePauw University, No. 8, Vol. 3 Part 1, March 1976; http://www.depauw.edu/sfs/backissues/8/christianson8art.htm.

214 **fast-forward through it rapidly:** For the author's own retelling of this wonderful story, see Ferguson, *Tycho and Kepler*.

215 *Mysterium Cosmographicum*, **which appeared in 1597:** Astronomer and historian of science Owen Gingerich has commented "Seldom in history has so wrong a book been so seminal in directing the future course of science." Owen Gingerich, "Johannes Kepler," in *Dictionary of Scientific Biography*, ed. Charles Coulston Gillespie, 7:289–312 (New York, NY: Charles Scribner's Sons, 1973), p. 292.

217 **summer's pastime or an intellectual exercise:** Christianson, "Kepler's *Somnium*."

218 **how they might be solved:** Ibid.

219 **"partly by history":** Johannes Kepler, *Joannis Kepleri Astronomi Opera Omnia, 8 Volumes*, Vol. 8, edited by Christian Frisch (Frankfurt-Erlangen: Heyder & Zimmer, 1858–1871), quoted in translation from John Lear, *Kepler's Dream, with the full text and notes of Somnium, Sive Astronomia Lunaris, Joannis Kepleri*, translated by Paricia Frueh Kirkwood (Los Angeles, CA: University of California Press, 1965), footnote to p. 12.

219 **"ancients called Thule":** Ibid., p. 22.

221 **"very wise spirits":** Ibid., p. 25.

221 **"superstition attached to them":** Kepler's *Somnium*, Kirkwood translation in Lear, Note 8.

221 "on goats or pitchforks": Lear, p. 26.

222 for his mother's witch trial: Ibid., Note 60.

223 "If I am not mistaken . . .": Ibid., Note 8.

223 "tutors in morals and studies": Ibid.

224 "Ignorance will finally die": Ibid., Note 2.

224 "still alive in the Universities": Ibid.

226 his own personal story and that of his mother: Lear, *Kepler's Dream*, p. 17.

226 being associated with "forbidden arts": Max Caspar, *Kepler*, translated and edited by C. Doris Hellman (London, England, and New York, NY: Abelard Schuman, 1959), p. 245.

226 "witch's antidote": Ferguson, p. 334.

227 "the right impression": Caspar, *Kepler*, p. 246.

227 "strenuous examination": See Caspar, *Kepler*, p. 255.

229 defend his mother himself: Caspar, *Kepler*, p. 252.

231 "not take His Holy Spirit from her but would stand by her": Frisch, vol. 8, pp. 549–550.

232 "in notes, ordered and numbered": Quoted in Carole Baumgardt, *Johannes Kepler: Life and Letters* (New York, NY: Citadel Press, 1953), p. 155.

232 "Table work": Kepler to Matthias Bernegger, February 8, 1627.

233 "translated from the Greek by me": Quoted in Edward Rosen, *Kepler's Somnium: The Dream, or Posthumous Work on Lunar Astronomy* (Madison, WI: University of Wisconsin Press, 1967); later published by Citadel Press (New York, NY, 1993). I used the earlier edition.

235 "instead I have made it my goal": Letter from Ludwig Kepler to Philip, Landgrave of Hesse, to whom Ludwig dedicated his work to complete the publication of *Somnium*. Caspar, *Kepler*, p. 364.

235 one unpublished and the other three out of print: They are: Lear, *Kepler's Dream*; Rosen, *Kepler's Somnium*; Joseph Keith Lane, unpublished manuscript on file at Columbia University's Carpenter Library under the title "The Dream; or Posthumous Work on Lunar Astronomy by Johannes Kepler"; and Everett F. Bleiler, published in a science fiction anthology, *Beyond Time and Space*, selected and with an introduction by Austust Derleth (New York, NY: Pelegrine and Cudahy, 1950).

235 and so did H. G. Wells: See Christianson, "Kepler's *Somnium*."

CHAPTER 9: ASTRONOMY ON ICE

237 "compensation from posterity—with interest." Milutin Milanković, *Milutin Milanković 1879-1958.* Autobiography with comments by his son, Vasko, and a preface by André Berger. (Katlenburg-Lindau, FRG: European Geophysical Society, 1995), p. 139.

238 "my voyage through the universe": Quoted in John Imbrie and Katherine Palmer Imbrie, *Ice Ages: Solving the Mystery* (Cambridge, MA, and London, England: Harvard University Press, 1979), p. 102.

239 an ancient age of ice had been responsible: Louis Agassiz, "Upon Glaciers, Moraines and Erratic Blocks," presented at a meeting of the Swiss Society of Natural Sciences, Neuchâtel, Switzerland, July 1837.

240 the *precession* of the equinoxes: John Herschel had been the first to discuss this precession, in 1830. (Milanković, *Milutin Milanković*, p. vi.)

241 James Croll's graph illustrating changes in orbital eccentricity: Figure A is redrawn from James Croll, "On the Eccentricity of the Earth's Orbit, and its Physical Relations to the Glacial Epoch," *Philosophical Magazine* 33 (1867): 119–131, in Imbrie and Imbrie, Figure 18. Also see James Croll's "Diagram Representing the Variations of Eccentricity of the Earth's Orbit," in *Climate and Time in the Geological Relations: A Theory of Secular Changes of the Earth's Climate* (New York, NY: D. Appleton & Company, 1893), p. 313.

241 **a point near Polaris:** To identify Polaris, look at the Big Dipper. An imaginary straight line beginning with the two stars at the opposite edge of the bowl from the handle and stretching outward from the bowl will point almost exactly to Polaris.

244 **at the Vienna Technical High School:** Information about these years comes from Milanković *Milutin Milanković*, pp. 13ff.

244 **"his greatest love":** Ibid., p. 14.

245 **"make out of me a robot":** Ibid., p. 16.

246 **"I felt I had not done badly":** Ibid., p. 36.

247 **"I thought I had both":** Ibid.

247 **"the founding stone of my future work":** Ibid., p. 39.

247 THE ICE AGES: Ibid., pp. 39–40.

249 **until the end of the war:** Short quotations and information about the war years are from ibid., pp. 44–46.

251 **never reaches those extremes:** Imbrie and Imbrie, *Ice Ages*, p. 107.

251 **as these numbers suggest:** André Berger, M. F. Loutre, and Q. Z. Yin, "Astronomical Theory of Paleoclimates," *Encyclopedia of Quaternary Science, vol. 2* (2013): 138.

252 **within the various atmospheric layers:** Milanković, *Milutin Milanković*, p. 51.

252 **"describe just the end result":** Ibid.

252 **"a class of excellent students":** Ibid., p. 56.

253 **"can recharge his batteries":** Ibid., p. 103.

255 **Milankovic's graph showing the radiation curve for latitude sixty-five degrees north:** Ibid., p. 107, English words added by Imbrie and Imbrie.

255 **"tackled the problem and solved it":** Vladimir Köppen, letter to Milutin Milanković, quoted in Milanković, *Milutin Milanković*, p. 86.

255 **"Yes, with your calculations!":** Ibid.

256 **"astronomically proven and the problem solved":** Wolfgang Sörgel, communication to Milutin Milanković, quoted in Milanković, p. 109.

256 **"could not believe his eyes":** Milanković, *Milutin Milanković*, p. 109.

256 **over the past million years:** Imbrie and Imbrie, *Ice Ages*, p. 197.

257 **"a weakness of my character":** Milanković, *Milutin Milanković*, p. 138.

257 **"company throughout the war":** Milanković, *Milutin Milanković*, p. 139.

258 **"posterity being its high court":** Ibid., p. 158.

258 **notably John Erikson and Cesare Emiliani:** Erikson and Emiliani did not work together. They were often not in agreement about their methods and clashed in what geologists remember as the "Erikson–Emiliani debate" in 1965.

259 **correspondence with Milankovic's graph of radiation curves:** Imbrie and Imbrie, *Ice Ages*, p. 137.

259 **80,000 years ago, and the present:** Ibid., p. 143.

259 **should be a warm period:** Ibid.

260 **and 400,000 years:** André Berger, "Astronomical Frequencies in Paleoclimates," *Encyclopedia of Marine Geosciences* (Dordrecht: Springer Science + Business Media, 2013), DOI: 10.1007/978-94-007-6644-0_215-1.

260 **for 500,000 years:** Ibid., p. 168.

260 **amplify the effect of the cycles:** Ian D. Whyte, *Climatic Change and Human Society* (London, England: Arnold Press; and New York, NY: Halsted Press, 1995), pp. 33–37.

260 **results (such as the Tibetan uplift):** C. Emiliani, "Milanković Theory Verified," *Nature* 364 (1993): 583–584.

NOTES

CHAPTER 10: THE ELUSIVE QUALITY OF STILLNESS

263 **into mainstream science:** Siegfried Othmer, "EEG Biofeedback: The Old and the New," 1994, revised 2004, accessed online November 6, 2015. Please note that this article is no longer available online.

263 **eighth-grade level:** Information about Sterman's student experience comes from the author's conversation with Sterman, and from Jim Robbins, *A Symphony in the Brain* (New York, NY: Grove Press, 2008), Kindle digital edition location 625ff.

263 **"definition of hypothyroidism":** Barry Sterman, e-mail to author, February 26, 2016.

264 **"to work with human beings":** Quoted in Robbins, *A Symphony*, Kindle edition location 625, from interview with Sterman.

266 **"in the literature of the time":** Ibid., p. 39, from interview with Sterman.

266 **in the brains of these cats:** M. B. Sterman and L. M. Thompson, "Neurofeedback for Seizure Disorders: Origins, Mechanisms, and Best Practice," in *Clinical Neurotherapy: Application of Techniques for Treatment*. David S. Cantor and James R. Evans, editors (San Diego, CA: Elsevier, 2013), pp. 301–302.

267 **waves of electrical activity are sweeping across it:** Carl Zimmer, "100 Trillion Connections: New Efforts Probe and Map the Brain's Detailed Architecture," *Scientific American* (January 2011): 59–63; https://www.scientificamerican.com/article/100-trillion-connections/.

268 **"in the history of clinical neurology":** David Millet, "The Origins of EEG," address at the Seventh Annual Meeting of the International Society for the History of the Neurosciences (ISHN), June 3, 2002.

268 **the reality of mental telepathy:** During a cavalry training exercise, he was thrown from his mount into the path of a horse-drawn cannon. The driver stopped in time, and Berger was not seriously injured. His sister, at home many miles away, had a strong feeling he was in danger and insisted that their father send him a telegram.

270 **surface of the head:** Hans Berger, "On the Electroencephalogram in Man," *Archive of Psychiatry and Nervous Diseases,* 1929.

270 **In 1941, Berger committed suicide:** Information in this paragraph comes from Millet, "The Origins of EEG."

272 **the United States, Britain, and France:** H. R. Wiedemann, "The Pioneers of Pediatric Medicine," *European Journal of Pediatrics* 153 (October 1994): 705.

273 **"influenced by expectation":** Sterman and Thompson, *Neurofeedback for Seizure Disorders,* p. 315.

273 **"cats did it in two weeks":** Interview with author, February 12, 2016.

274 **thorough records of her condition:** Robbins, *A Symphony*, Kindle edition location 810.

274 **"sleep patterns and personality were noted also":** M. B. Sterman and L. Friar, "Suppression of seizures in an epileptic following sensorimotor EEG feedback training," *Electroencephalography and Clinical Neurophysiology* 33, No. 1 (July 1972): 89.

276 **in the journal *Epilepsia*:** M. B. Sterman, L. R. Macdonald, and R. K. Stone, "Biofeedback Training of the Sensorimotor Electroencephalogram Rhythm in Man: Effects on Epilepsy," *Epilepsia* (September 1974): 395–416.

276 **a 74 percent reduction:** M. B. Sterman and L. R. Macdonald, "Effects of Central Cortical EEG Feedback Training on Incidence of Poorly Controlled Seizures," *Epilepsia* (June 1978): 205–322.

278 **"pure politics":** Quoted in Robbins, *A Symphony*, Kindle edition location 926.

278 **"who had been on the committee!":** Barry Sterman, e-mail to author, February 26, 2016.

278 **"people's minds snap shut":** Quoted in Robbins, *A Symphony*, Kindle edition location 119, from an interview with Sterman in the preface to the revised and expanded 2008 edition.

279 *including* **the brain:** Neurofeedback is in fact one kind of EEG biofeedback. The reason the field coined the term "neurofeedback" was in part because the word biofeedback was becoming associated in the public's mind with "New Age-y" science.

280 **at Hollywood Presbyterian Hospital:** For this information on Sterman's life after his funding was cut: Robbins, *A Symphony,* Kindle edition location 927.

281 **"a piece of cake":** "The History of Neurofeedback - Sterman/Lubar Studies," *Brain and Body Solutions;* http://www.brainandbodysolutions.com/learn/how-neurofeedback-works/.

281 **raised their IQs ten to fifteen points:** Robbins, *A Symphony,* Kindle edition location 1947.

284 **occur very rapidly:** This explanation is taken broadly from "EEG Changes," *About Neurofeedback: Information, Perspective and Advice;* http://www.aboutneurofeedback.com/neurofeedback-info-center/faq/how-does-neurofeedback-work/eeg-changes/.

285 **either awake or asleep:** Sterman has provided the author with this new but documented proprietary information.

285 **"Sterman observed in his cats":** Barry Sterman, e-mail to author, March 28, 2016.

286 **"the stage is set for sleep":** Sterman to author, e-mail, March 28, 2016.

286 **as well as on the arousal level:** These paragraphs on the mechanism of SMR are paraphrased by the author from M. B. Sterman and L. M. Thompson, *Neurofeedback for Seizure Disorders,* pp. 302–303.

287 **individual's arousal level:** The image of a symphony orchestra and its conductor was suggested to the author by Dr. Joseph Sandford.

287 **"evidence of individual case histories":** Siegfried Othmer, "EEG Biofeedback." Please note that this article is no longer available online.

288 **"from [epileptic] seizures":** M. B. Sterman and L. M. Thompson, *Neurofeedback for Seizure Disorders,* p. 307

289 **children and adults with ADHD:** "The History of Neurofeedback – Sterman/Lubar Studies."

289 **"what our brain does well":** Siegfried Othmer, "EEG Biofeedback." Please note that this article is no longer available online.

BIBLIOGRAPHY

Agassiz, Louis. "Upon Glaciers, Moraines and Erratic Blocks." Presented at a meeting of the Swiss Society of Natural Sciences. Neuchâtel, Switzerland. July 1837.

Alic, Margaret. *Hypatia's Heritage: A History of Women in Science from Antiquity through the Nineteenth Century*. London, England: Women's Press Limited, 1986.

Anderson, Mark. *The Day the World Discovered the Sun: An Extraordinary Story of Scientific Adventure and the Race to Track the Transit of Venus*. Boston, MA: Da Capo Press, 2012.

Armitage, Angus. "Chappe D'Auteroche: A Pathfinder for Astronomy." *Annals of Science* 10, No. 4 (1954).

Badash, Laurence, editor. *Rutherford and Boltwood: Letters on Radioactivity*. New Haven, CT: Yale University Press, 1969.

Berger, André. "Astronomical Frequencies in Paleoclimates." *Encyclopedia of Marine Geosciences*. Dordrecht: Springer Science + Business Media, 2013. DOI: 10.1007/978-94-007-6644-0_215-1.

———, M. F. Loutre, and Q. Z. Yin. "Astronomical Theory of Paleoclimates." *Encyclopedia of Quaternary Science* 2 (2013): 138.

Berger, Hans. "On the Electroencephalogram in Man." *Archive of Psychiatry and Nervous Diseases* (1929).

Berthelet, M., and Cu. Em. Ruelle. *Collection des Anciens Alchimistes Grecs*. Paris: Georges Steinhei, 1888.

Bleiler, Everett F. In *Beyond Time and Space*. New York, NY: Pelegrine and Cudahy, 1950.

Brafman, David, and Stephanie Schrader. *Insects and Flowers: The Art of Maria Sibylla Merian*. Los Angeles, CA: J. Paul Getty Museum, 2008.

Brown, G. I. *Scientist, Soldier, Statesman, Spy: Count Rumford – The Extraordinary Life of a Scientific Genius*. Phoenix Mill, England: Sutton Publishers, 1999.

Brown, Sanborn C. *Benjamin Thompson, Count Rumford*. Cambridge, MA: MIT Press, 1979.

———, "Count Rumford and the Caloric Theory of Heat." *Proceedings of the American Philosophical Society* 93, No. 4: The Theory of Relativity in Contemporary Science (1949): 324; http://www.jstor.org/stable/3143157.

———, editor. *The Collected Works of Count Rumford, Volume I: The Nature of Heat*. Cambridge, MA: The Belknap Press of Harvard University Press, 1968.

———, *The Collected Works of Count Rumford, Volume III: Devices and Utensils*. Cambridge, MA: The Belknap Press of Harvard University Press, 1969, p. 77.

———, *The Collected Works of Count Rumford, Volume V: Public Institutions*. Cambridge, MA: The Belknap Press of Harvard University Press, 1970.

Baumgardt, Carole. *Johannes Kepler: Life and Letters*. New York, NY: Citadel Press, 1953.

Bell, Thomas. "Presidential Address to the Members of the Linnean Society." *Journal of the Linnean Society (Zoology)* 4 (1859): viii–ix.

Brackman, Arnold. *A Delicate Arrangement: The Strange Case of Charles Darwin and Alfred Russel Wallace*. New York, NY: Times Books, 1980.

Broda, Engelbert. *Ludwig Boltzmann: Mensch, Physiker, Philosoph*. Wien: Franz Deuticke, 1986.

Bronowski, Jacob. *The Ascent of Man*. Boston, MA: Little, Brown & Co., 1973.

Brooks, John Langdon. *Just Before the Origin: Alfred Russel Wallace's Theory of Evolution*. New York, NY: Columbia University Press, 1984.

Burkert, Walter. *Lore and Science in Ancient Pythagoreanism*. Translated by Edwin L. Minar, Jr. Cambridge, MA: Harvard University Press, 1972.

Caspar, Max. *Kepler*. Translated and edited by C. Doris Hellman. London and New York : Abelard Schuman, 1959.

———, Walther von Dyck, Franz Hammer, and Volker Bialas, eds. *Johannes Kepler Gesammelte Werke*, 22 vols. Munich: Deutsche Forschungsgemeinschaft, and the Bavarian Academy of Sciences, 1937–.

Christianson, Gale E. "Kepler's *Somnium:* Science Fiction and the Renaissance Scientist." *Science Fiction Studies, DePauw University.* URL: http://www.depauw.edu/sfs/backissues/8/christianson8art.htm.

Count of Rumford, Benjamin. "An Inquiry concerning the Source of the Heat Which is Excited by Friction," Philosophical Transaction of the Royal Society of London 88 (January 1798): 80–102. DOI: 10.1098/rstl.1798.0006 Phil. Trans. R. Soc. Lond.

Croll, James. "On the Eccentricity of the Earth's Orbit, and its Physical Relations to the Glacial Epoch." *Philosophical Magazine 33* (1867): 119–131.

——, "Diagram Representing the Variations of Eccentricity of the Earth's Orbit." In *Climate and Time in the Geological Relations: A Theory of Secular Changes of the Earth's Climate.* New York, NY: D. Appleton & Company, 1893.

Darwin, Charles. Letter to Charles Lyell. June 18, 1858. Record number WCP5647. "Wallace Letters Online." *Natural History Museum.* URL: http://www.nhm.ac.uk/research-curation/scientific-resources/collections/library-collections/wallace-letters-online/5647/6498/details.html. Record created August 18, 2014.

——, Letter to Charles Lyell. July 18, 1858. Record number WCP5651. "Wallace Letters Online." *Natural History Museum.* URL: http://www.nhm.ac.uk/research-curation/scientific-resources/collections/library-collections/wallace-letters-online/5651/6502/details.html. Record created August 18, 2014.

——, Letter to Charles Lyell. June 25, 1858. Record number WCP5648. "Wallace Letters Online." *Natural History Museum.* URL: http://www.nhm.ac.uk/research-curation/scientific-resources/collections/library-collections/wallace-letters-online/5648/6499/details.html. Record created August 18, 2014.

——, Letter to Alfred Russel Wallace. January 25, 1859. Record number WCP1841. "Wallace Letters Online." *Natural History Museum.* URL: http://www.nhm.ac.uk/research-curation/scientific-resources/collections/library-collections/wallace-letters-online/1841/5924/details.html. Record created June 7, 2013.

——, Letter to Alfred Russel Wallace. April 6, 1859. Record number WCP1842. "Wallace Letters Online." *Natural History Museum.* URL: http://www.nhm.ac.uk/research-curation/scientific-resources/collections/library-collections/wallace-letters-online/1842/5925/details.html. Record created June 7, 2013.

——, Letter to Alfred Russel Wallace. November 13, 1859. Record number WCP1844. "Wallace Letters Online." *Natural History Museum.* URL: http://www.nhm.ac.uk/research-curation/scientific-resources/collections/library-collections/wallace-letters-online/1844/5927/details.html. Record created June 7, 2013.

——, Letter to Alfred Russel Wallace. May 18, 1860. Record number WCP1846. "Wallace Letters Online." *Natural History Museum.* URL: http://www.nhm.ac.uk/research-curation/scientific-resources/collections/library-collections/wallace-letters-online/1846/5929/details.html. Record created June 7, 2013.

Darwin Correspondence Project. "Alfred Russel Wallace's essay on varieties." *University of Cambridge.* URL: https://www.darwinproject.ac.uk/people/about-darwin/origin-species/alfred-russel-wallace-s-essay-varieties.

Davis, Natalie Zemon. *Women on the Margins: Three Seventeenth-Century Lives.* Cambridge, MA: Harvard University Press, 1995.

d'Auteroche, Chappe. *A Journey into Siberia, Made by Order of the King of France.* London, England: T. Jefferys, 1770. URL: http://hdl.handle.net/2027/uc2.ark:/13960/t1dj5pz7d.

d'Auteroche, Chappe. *A Voyage to California: to Observe the Transit of Venus* (London, England: E. and C. Dilly, 1773), p. 2. URL: http://hdl.handle.net/2027/nyp.33433000631253.

Dean, Thatcher E. "The Chinese Imperial Astronomical Bureau: Form and Function of the Ming Dynasty 'Qintianjian' from 1365 to 1627." PhD Thesis. Seattle, WA: University of Washington, 1989.

de Labadie, Jean. *Les Entretiens d'Esprit du Jour Chretien, ou les Reflexions Impotantes du Fidele.* Amsterdam: Laurans Autein, 1671.

Drummond, William. "On the Science of the Egyptians and Chaldeans." *The Classical Journal*, Vol. 18. London, England: A. J. Valpy, 1818.

"EEG Changes." *About Neurofeedback: Information, Perspective and Advice.* URL: http://www.aboutneurofeedback.com/neurofeedback-info-center/faq/how-does-neurofeedback-work/eeg-changes/.

Editors of *Encyclopædia Britannica*, online, s.v. "Adam Schall von Bell." *Encyclopædia Britannica*. Chicago: Encyclopædia Britannica, Inc., 2008. URL: http://www.britannica.com/EBchecked/topic/527026/Adam-Schall-von-Bell?anchor=ref1003343.

———, "Kangxi: Emperor of Qing Dynasty." *Encyclopædia Britannica*. Chicago: Encyclopædia Britannica, Inc., 2008. URL: http://www.britannica.com/biography/Kangxi.

Einstein, Albert. *Out of My Later Years*. New York, NY: Philosophical Library, 1950.

———, "The Development of Our Views on the Nature and Constitution of Radiation." Salzburg, Austria: 1909.

Ellis, George E. *Memoir of Sir Benjamin Thompson, Count Rumford, with Notices of His Daughter*. Boston: American Academy of Arts and Sciences, 1871.

Elman, Benjamin A. *On Their Own Terms: Science in China 1550–1900*. Cambridge, MA, and London, England: Harvard University Press, 2008.

"Éloge de M. l'Abbé Chappe." In *Histoire de L'Académie Royale des Sciences*. Paris: 1772.

Emiliani, C. "Milanković Theory Verified." *Nature* 364 (1993): 583–584.

Ferguson, Kitty. *The Music of Pythagoras: How an Ancient Brotherhood Cracked the Code of the Universe and Lit the Path from Antiquity to Outer Space*. New York, NY: Walker Books, 2008.

———, *Tycho & Kepler: The Unlikely Partnership That Forever Changed Our Understanding of the Heavens*. New York, NY: Walker & Company, 2002.

Frisch, Otto. "The Discovery of Fission: How It All Began." *Physics Today* 20, No. 11 (November 1967): 47.

———, *What Little I Remember*, Cambridge, England: Cambridge University Press, 1979.

Gingerich, Owen. "Johannes Kepler." *Dictionary of Scientific Biography*. Edited by Charles Coulston Gillespie. New York, NY: Charles Scribner's Sons, 1973.

Golvers, Noël. *Ferdinand Verbiest, S. J. (1623–1688) and the Chinese Heaven: The Composition of his Astronomical Corpus and its Reception in the European Republic of Letters (Louvain Chinese Studies)*. Belgium: Leuven University Press, 2003.

———, *The Astronomia Europaea of Ferdinand Verbiest, S. J. (Dillingen, 1687): Text, Translation, Notes and Commentaries*. Nettetal, Germany: Steyler Verlag, 1993.

———, and Efthymios Nicolaidis. *Ferdinand Verbiest and Jesuit Science in 17th Century China: An Annotated Edition and Translation of the Constantinople Manuscript (1676)*. Athens, Greece/Leuven, Belgium: Institute for Neohellenic Research/Ferdinand Verbiest Institute, 2009.

Guthrie, W. K. C. *A History of Greek Philosophy, Vol I: The Earlier Presocratics and the Pythagoreans*. Cambridge, England: Cambridge University Press, 1962.

Hahn, Otto. Letter to Lise Meitner. December 19, 1938. "Meitner Collection." *Churchill College Archives Center*. Cambridge, England. MTNR 5/21A.

———, *Mein Leben*. Translated by E. Kaiser and E. Wilkins as *Otto Hahn: My Life*. London, England: MacDonald & Co., 1970.

———, and Lise Meitner. "Die Muttersubstanz des Actiniums." *Physikalische Zeitschrift* 19 (1918): 208–218.

Hecht, Eugene. *Physics in Perspective*. New York, NY: Addison-Wesley, 1962.

Imbrie, John, and Katherine Palmer Imbrie. *Ice Ages: Solving the Mystery*. Cambridge, MA, and London, England: Harvard University Press, 1979.

Josson, H., and L. Willaert, editors. *Correspondance de Ferdinand Verbiest de la Compagnie de Jésus (1623–1688): Directeur de l'observatoire de Pékin*. Brussels: 1938.

Joule, James Prescott. "On the Mechanical Equivalent of Heat." *Philosophical Transaction of the Royal Society of London* 140 (January 1850): 61–82. DOI: 10.1098/rstl.1850.0004 Phil. Trans. R. Soc. Lond.

Kepler, Johannes. *Joannis Kepleri Astronomi Opera Omnia, 8 Volumes.* Vol. 8. Edited by Christian Frisch. Frankfurt-Erlangen: Heyder & Zimmer, 1858–1871.

Kepler's Somnium. Translated by Edward Rosen. Madison, WI: University of Wisconsin Press, 1967.

Kerner, Charlotte. *Lise, Atomphysikerin: Die Lebensgeschichte der Lise Meitner.* Wernheim and Basel: Beltz and Gelberg, 1987.

Kramish, Arnold. *The Griffin: The Greatest Untold Spy Story of World War II.* Boston, MA: Houghton Mifflin Company, 1986.

Lane, Joseph Keith. Unpublished manuscript under the title "The Dream; or Posthumous Work on Lunar Astronomy by Johannes Kepler." On file at Columbia University's Carpenter Library.

Lear, John. *Kepler's Dream, with the full text and notes of Somnium, Sive Astronomia Lunaris, Joannis Kepleri.* Translated by Patricia Frueh Kirkwood. Los Angeles, CA: University of California Press, 1965.

Lemmerich, Jost, editor. *Die Geschichte der Entdeckung der Kernspaltung: Ausstellungskatalog.* Berlin: Technische Universität Berlin, Universitätsbibliothek, 1988.

Lienhard, John H. "Engines of Our Ingenuity No. 4: Count Rumford." *University of Houston.* URL: http://www.uh.edu/engines/epi4.htm.

"Lise Meitner, A physicist who never lost her humanity." Powerpoint presentation. *Department of Physics & Astronomy, The University of Utah.* http://www.physics.utah.edu/~jui/3375/Class%20Materials%20Files/y2007m09d24/LiseMeitner.pdf.

Lindel, Robert. "Music and Patronage at the Court of Rudolf II." In *Music in the German Renaissance: Sources, Styles, and Contexts.* Edited by John Kmetz. Cambridge, England: Cambridge University Press, 1994.

MacDonnell, Joseph F. "Fr. Ferdinand Verbiest, S.J. (1623–1688) a Jesuit Scientist in China." *Fairfield University Mathematics Department.* http://www.faculty.fairfield.edu/jmac/sj/scientists/verbiest.htm.

Merian, Maria Sibylla. *Metamorphosis insectorum Surinamensium.* In *Maria Sibylla Merian in Surinam, Kommentar zur Faksimile-Ausgabe.* Translated by Elisabeth Rücker and William T. Stearn. London, England: Pion, 1982.

———, *The Wondrous Transformation of Caterpillars: Fifty Engravings Selected from Erucarum Ortus (1718).* Introduction by William T. Stern. London, England: Scolar Press, 1978.

———, "Einige Erinnerungen an das Kaiser-Wilhelm-Institut für Chemie in Berlin-Dahlen." *Naturwissenschaften* 41 (1954): 970–999.

———, Letter to Fräulein Hitzenberger. April 10 and 29, 1951. "Meitner Collection." *Churchill College Archives Center.* Cambridge, England.

———, Letter to Otto Hahn. October 14, 1915. The literary estate of Otto Hahn. *Max-Planck-Gesellschaft Archives.* Berlin-Dahlem, Germany.

———, Letter to Otto Hahn. August 24, 1938. "Meitner Collection." *Churchill College Archives Center.* Cambridge, England. MTNR 5/21A.

———, Letter to Otto Hahn. March 10, 1939. "Meitner Collection." *Churchill College Archives Center.* Cambridge, England. MTNR 5/21B.

———, Letter to Paul Rosbaud. August 4, 1946. "Meitner Collection." *Churchill College Archives Center.* Cambridge, England. MTNR 5/15. Document No. 44.

———, "Status of Women in the Professions." *Physics Today* (August 1960): 20.

———, "Wege und Irrwege zur Kernenergie." *Naturwissenschaften Rund* 16 (May 1963): 167.

———, "Looking Back." *Bulletin of the Atomic Scientists* 20 (November 1964): 2.

———, and Otto R. Frisch. "Disintegration of Uranium by Neutrons: A New Type of Nuclear Reaction," *Nature* 143 (February 1939): 239–240.

Mignard, F. "The Solar parallax with the transit of Venus," Version 4.1. Thesis, Observatoire de la Côte d'Azur, 2004. URL: https://www-n.oca.eu/Mignard/Transits/Data/venus_contact.pdf.

Milanković, Milutin. *Milutin Milanković 1879-1958*. Katlenburg-Lindau, FRG: European Geophysical Society, 1995.

Millet, David. "The Origins of EEG." Address at the seventh annual meeting of the International Society for the History of the Neurosciences (ISHN). June 3, 2002.

Needham, Joseph. *Chinese Astronomy and the Jesuit Mission: An Encounter of Cultures*. London, England: The China Society, 1958.

Nunis, Doyce B., Jr., editor. *The 1769 Transit of Venus: The Baja California Observations of Jean-Baptiste Chappe d'Auteroche, Vicente de Doz, and Joaquín Velázquez Cárdenas de León*. Translation by Iris H. Engstrand, Maynard J. Geiger, and James Donahue. Los Angeles, CA: Natural History Museum of Los Angeles County, 1982.

Owens, Susan. "Maria Sibylla Merian: 'Great Diligence, Grace and Spirit.'" In *Amazing Rare Things: The Art of Natural History in the Age of Discovery*. Edited by David Attenborough. New Haven, CT, and London, England: Yale University Press, 2007.

Patai, Raphael. *The Jewish Alchemists: A History and Source Book*. Princeton, NJ: Princeton University Press, 1994 and 2014.

Pauli, Wolfgang. Telegram to Dirk Coster. July, 1938.

Pindar, Peter. *The Works of Peter Pindar, Esq. to which are prefixed Memoirs of the Author's Life*. Volume II. London, England: 1802.

Prime, Nathaniel S. *A History of Long Island: From its First Settlement by Europeans to the year 1845, with Special Reference to its Ecclesiastical Concerns*. New York, NY: Robert Carter, 1845.

Richardson, Kenneth. *The British Motor Industry 1896–1939: A Social & Economic History*. London, England: Macmillan, 1977.

Rife, Patricia. *Lise Meitner and the Dawn of the Nuclear Age*. Boston, MA: Birkhäuser, 1999.

Robbins, Jim. *A Symphony in the Brain*. New York, NY: Grove Press, 2008.

Rosen, Edward. *Kepler's Somnium: The Dream, or Posthumous Work on Lunar Astronomy*. Madison, WI: University of Wisconsin Press, 1967.

Rouleau, Francis A., and Edward J. Malatesta. "The 'Excommunication' of Ferdinand Verbiest." In *Ferdinand Verbiest, S. J. (1623–1688): Jesuit Missionary, Scientist, Engineer and Diplomat*. Edited by John W. Witek, S. J. Leuven, Belgium: Institut Monumenta Serica, 1994.

Rowlinson, Hugh. "The Contribution of Count Rumford to Domestic Life in Jane Austen's Time." *Jane Austen Society of North America: Persuasions On-Line* 23, No. 1 (Winter 2002). URL: http://www.jasna.org/persuasions/on-line/vol23no1/rowlinson.html.

Rumford, Benjamin Graf von. *Mémoires sur la Chaleur, par Le Comte de Rumford, V. P. R. S.* Paris: Associé Étranger de L'Institut National, 1804.

Schiebinger, Londa. *The Mind Has No Sex?: Women in the Origins of Modern Science*. Boston, MA, and London, England: Harvard University Press, 1989.

Schürer, Emil. "Alexandria, Egypt—Ancient." *Jewish Encyclopedia: The Unedited Full-Text of the 1906 Jewish Encyclopedia*. URL: http://www.jewishencyclopedia.com/articles/1171-alexandria-egypt-ancient.

Sime, Ruth Lewin. *Lise Meitner: A Life in Physics*. Berkeley, CA: University of California Press, 1996.

Slotten, Ross A. *The Heretic in Darwin's Court: The Life of Alfred Russel Wallace*. New York, NY: Columbia University Press, 2004.

Standaert, Nicolas. *The Interweaving of Rituals: Funerals in the Cultural Exchange between China and Europe*. Seattle, WA: University of Washington Press, 2008.

Sterman, Barry M. E-mail to author, February 26, 2016.

——, and L. Friar. "Suppression of seizures in an epileptic following sensorimotor EEG feedback training." *Electroencephalography and Clinical Neurophysiology* 33, No. 1 (July 1972): 89.

——, and L. M. Thompson. "Neurofeedback for Seizure Disorders: Origins, Mechanisms, and Best Practice." In *Clinical Neurotherapy: Application of Techniques for Treatment*. Edited by David S. Cantor and James R. Evans. San Diego, CA: Elsevier, 2013.

———, and L. R. Macdonald. "Effects of Central Cortical EEG Feedback Training on Incidence of Poorly Controlled Seizures." *Epilepsia* (June 1978): 205–322.

———, L. R. Macdonald, and R. K. Stone. "Biofeedback Training of the Sensorimotor Electroencephalogram Rhythm in Man: Effects on Epilepsy." *Epilepsia* (September 1974): 395–416.

Surburg, Raymond F. *Introduction to the Intertestamental Period*. St. Louis, MO: Concordia Publishing House, 1975.

Taylor, F. Sherwood. "The Evolution of the Still." *Annals of Science 5* (1945): 190.

Taylor, Joan E. *Jewish Women Philosophers of First-Century Alexandria: Philo's 'Therapeutae' Reconsidered*. Oxford, England: Oxford University Press, 2003.

"The History of Neurofeedback - Sterman/Lubar Studies," *Brain and Body Solutions*, URL: http://www.brainandbodysolutions.com/learn/how-neurofeedback-works/.

Thompson, Colonel Sir Benjamin. In a letter to Sir Joseph Banks. "New Experiments upon Heat." *Philosophical Transactions of the Royal Society of London* 76 (January 1786): 273–304. DOI: 10.1098/rstl.1786.0014 Phil. Trans. R. Soc. Lond.

Todd, Kim. *Chrysalis: Maria Sibylla Merian and the Secrets of Metamorphosis*. New York, NY: Harcourt, 2007.

Wallace, Alfred Russel. Acceptance speech. Printed in *The Darwin-Wallace Celebration*. Thursday, July 1, 1908. London, England: Printed for the Linnean Society, Burlington House, Piccadilly, W, 1908.

———, *A Narrative of Travels on the Amazon and Rio Negro, With an Account of the Native Tribes, and Observations on the Climate, Geology, and Natural History of the Amazon Valley*. 2nd edition. London, England; New York, NY; and Melbourne, Australia: Ward, Lock and Co., Minerva Library of Famous Books, 1889. URL: https://archive.org/details/travelsonamazonr00wall.

———, "Equatorial Vegetation." In *"Tropical Nature" and Other Essays, 1878*. Reprint. New York, NY: AMS Press, 1975.

———, *My Life: A Record of Events and Opinions*, Volumes 1 and 2. New York, NY: Dodd, Mead & Company, 1906.

———, Letter to Charles Darwin. September 27, 1857. Record number WCP4080. "Wallace Letters Online." *Natural History Museum*. URL: http://www.nhm.ac.uk/research-curation/scientific-resources/collections/library-collections/wallace-letters-online/4080/4027/details.html. Record created March 8, 2012.

———, Letter to George Silk. November 30, 1858. Record number WCP370. "Wallace Letters Online." *Natural History Museum*. URL: http://www.nhm.ac.uk/research-curation/scientific-resources/collections/library-collections/wallace-letters-online/370/370/details.html. Record created June 1, 2002.

———, Letter to George Silk. September 1, 1860. Record number WCP363. "Wallace Letters Online." *Natural History Museum*. URL: http://www.nhm.ac.uk/research-curation/scientific-resources/collections/library-collections/wallace-letters-online/363/363/details.html. Record created June 1, 2002.

———, Letter to Henry Bates. December 24, 1860. Record number WCP374. "Wallace Letters Online." *Natural History Museum*. URL: http://www.nhm.ac.uk/research-curation/scientific-resources/collections/library-collections/wallace-letters-online/374/5916/details.html. Record created June 3, 2013.

———, Letter to Joseph Dalton Hooker. October 6, 1858. Record number WCP1454. "Wallace Letters Online." *Natural History Museum*. URL: http://www.nhm.ac.uk/research-curation/scientific-resources/collections/library-collections/wallace-letters-online/1454/4022/details.html. Record created March 7, 2012.

———, Letter to Mary Anne Wallace. October 6, 1858. Record number WCP369. "Wallace Letters Online," *Natural History Museum*, URL: http://www.nhm.ac.uk/research-curation/scientific-resources/collections/library-collections/wallace-letters-online/369/5914/details.html. Record created June 3, 3013.

———, *On Miracles and Modern Spiritualism: Three Essays*. London, England: James Burns, 1875.

———, "On the Entomology of the Aru Islands." *Zoologist* 16 (1858): 5889–5894.

——, "On the Law which has Regulated the Introduction of New Species." *Annals and Magazine of Natural History* 16 (1855): 184–196. URL: http://www.esp.org/books/wallace/law.pdf.

——, "On the Tendency of Varieties to Depart Indefinitely from the Original Type; Instability of Varieties Supposed to Prove the Permanent Distinctness of Species." *Journal of the Linnean Society (Zoology)* 3 (1858): 53–62.

——, "On the Umbrella Bird (Cephalopterus ornatus), 'Ueramimbé.'" *Annals and Magazine of Natural History* Vol. 8, 47 (1851): 428–430. DOI: 10.1080/03745486109494996. Published online December 23, 2009.

——, "The Limits of Natural Selection as Applied to Man." In *Contributions to the Theory of Natural Selection: A Series of Essays*. London, England: Macmillan, 1870.

——, *The Malay Archipelago*. Reprinted edition. New York, NY: Oxford University Press, 1986.

——, "The Ornithology of Northern Celebes." *IBIS: International Journal of Avian Science* 2, No. 2 (April 1860): 140–147. DOI: 10.1111/j.1474-919X.1860.tb06361.x.

von Hevesy, George. Letter to Ernest Rutherford. October 14, 1913. Rutherford Papers. Cambridge, England: Cambridge University.

Walle, Willy Vande. "Ferdinand Verbiest and the Chinese Bureaucracy." In *Ferdinand Verbiest, S. J. (1623–1688): Jesuit Missionary, Scientist, Engineer and Diplomat*. Edited by John W. Witek, S. J. Leuven, Belgium: Institut Monumenta Serica, 1994.

Ward, George Atkinson. *Journal and Letters of the Late Samuel Curwen, Judge of Admiralty: An American Refugee in England, from 1775 – 1784, Comprising Remarks on the Prominent Men and Measures of that Period, to which are added Biographical Notices of Many American Loyalists, and Other Eminent Persons*. New York, NY: C. S. Francis, 1842.

Weld, Charles. *A History of the Royal Society: With Memoirs of the Presidents*, Vol. II. Cambridge, MA: Cambridge University Press, 2011.

Whyte, Ian D. *Climatic Change and Human Society*. London, England: Arnold Press; and New York, NY: Halsted Press, 1995.

Wiedemann, H.R. "The Pioneers of Pediatric Medicine." *European Journal of Pediatrics* 153 (October 1994): 705.

Witek, John W., editor. *Ferdinand Verbiest, S. J. (1623–1688): Jesuit Missionary, Scientist, Engineer and Diplomat*. Leuven, Belgium: Institut Monumenta Serica, 1994.

Wulf, Andrea. *Chasing Venus: The Race to Measure the Heavens*. New York, NY: Vintage Books, 2013.

Youmans, Edward L. *The Correlation and Conservation of Forces: A Series of Expositions, by Prof. Grove, Prof. Helmholtz, Prof. Liebig and Dr. Carpenter. With an Introduction and Brief Biographical Notices of the Chief Promoters of the New Views*. New York, NY: D. Appleton and Company, 1864.

Zimmer, Carl. "100 Trillion Connections: New Efforts Probe and Map the Brain's Detailed Architecture." *Scientific American* (January 2011): 59–63; https://www.scientificamerican.com/article/100-trillion-connections/.

INDEX

INDEX

IMAGE CREDITS

Τοῦ ἄνω σφεραδ ἕξεως στ̣α̣ ἔχοντας κ̣ ἐπ̣ι̣θεν̣ι̣
χαλκ̣ου ἐπι̣το̣υλω̣λ̣ω̣ πε̣αδ̣ος ὀστράκι̣νης ἐχου̣σ̣
Τὸ ὀφ̣αριου σ̣τ̣ηρ̣ιτη̣λῶ̣ σωτ̣α̣ο σι̣μβ̣ολα̣ω̣ ε̣νδε̣
ἐστ̣α ἄκρ̣ε τῶ γ̣ουλη̣υ̣β̣ηκου̣ν̣ υ̣δ̣χ̣ι̣γ̣ο̣ι̣ο̣
μ̣ῆ̣ ἄχοι̣σ̣ παχυς̣ι̣γ̣εμ̣ερ̣αγ̣ω̣ο̣ ναπο̣γ̣η̣δ̣
θερμ̣ας̣τ̣οῦ ὕδατος κομ̣ιζ̣ούσ̣ης̣ Τὸ α̣γ̣χ̣ε̣ι̣μ̣ου̣·Τὸ
δ̣ε̣ τ̣χ̣ημα̣ Τοῦτο̣ λ̣ι̣χαγ̣ο σ̣ου̣λ̣η·– Ἑστ̣ι̣ δε̣χ̣η̣
λος̣ Ρο̣τ̣ε̣σ̣κομ̣ιδ̣ηευ̣ δ̣υ̣τ̣ος εθου·εχ̣λου̣χ̣ω̣ετ̣ι̣
βηκος·εστ̣ω̣σ̣ου̣λη̣γ̣ε̣τ̣ω̣θ̣η̣λ̣ι̣να̣ χαλκι̣ου εν̣τ̣εθ̣η̣
μ̣ι̣ν̣ βηκος τ̣ή̣τ̣ε̣ος ο̣ς ϛ̄·Τ̣ῶ̣αντῶ̣ ρ̣ο̣τ̣ε̣κ̣ἡ̣βη̣
κος φ̣ε̣ κ̣η̣υ̣ποκ̣ατ̣ω̣λ̣ω̣τ̣α̣ο̣θ̣φ̣ουα̣πυ̣ρου·φ̣ε̣η̣υ̣
σ̣ι̣ω̣φρμα̣ζ̣ο̣τ̣ο̣χ̣α̣λ̣κ̣ι̣ο̣γ̣κ̣η̣ζ̣δ̣ε̣ι̣ τ̣η̣λ̣ο̣ι̣σ̣τ̣ε̣υ̣τ̣ῆ̣
χηρ̣ῶ̣ η̣ στ̣η̣λ̣ω̣·η̣ω̣ε̣β̣ου̣λ̣η̣κ̣η̣κ̣ω̣ι̣ο̣ν̣ο̣α̣μ̣α̣τ̣α̣ι̣·
ο̣ι̣ δε̣ Τ̣υ̣π̣ε̣ι̣ ου̣τ̣ο̣ι·⁓

XLII.

APPARATUS *adapted*

for shewing The